Environmental Science and Engineering

Environmental Science and Engineering

Carol Thorburn

Larsen & Keller
www.larsen-keller.com

Environmental Science and Engineering
Carol Thorburn
ISBN: 978-1-64172-092-2 (Hardback)

Larsen & Keller

Published by Larsen and Keller Education,
5 Penn Plaza,
19th Floor,
New York, NY 10001, USA

Cataloging-in-Publication Data

Environmental science and engineering / Carol Thorburn.
    p. cm.
Includes bibliographical references and index.
ISBN 978-1-64172-092-2
1. Environmental sciences. 2. Environmental engineering. 3. Environmental management.
4. Environmental protection. I. Thorburn, Carol.
GE105 .E58 2019
363.7--dc23

For more information regarding Larsen and Keller Education and its products, please visit the publisher's website www.larsen-keller.com

Table of Contents

Preface

Environmental science is the academic domain concerned with the integrated study of environmental systems. The study of environmental issues and the formulation of strategies are also under the scope of its study. An understanding of atmospheric sciences, environmental chemistry, ecology and geosciences is intrinsic to environmental science. There has been significant human intervention in the ecology and biodiversity of the Earth, which has resulted in adverse environmental conditions. Therefore, it becomes imperative to develop and implement effective strategies for protecting human populations and the natural environment from these effects and improve environmental quality. This may be achieved by an integration of scientific and engineering principles which is undertaken by the field of environmental engineering. Waste disposal and management, environmental sustainability, pollution control and management, etc. are some of the methods that can be used in the management of environmental problems. This book is a valuable compilation of topics, ranging from the basic to the most complex theories and principles in the fields of environmental science and engineering. Different approaches, evaluations and methodologies on these fields have been included herein. This textbook, with its detailed analyses and data, will prove immensely beneficial to professionals and students involved in this area at various levels.

Given below is the chapter wise description of the book:

Chapter 1, Environmental science is an academic field that integrates biological, physical and information sciences for the study of the environment. A branch of engineering that is concerned with the protection of human populations from the effects of adverse environmental conditions is under the domain of environmental engineering. This chapter has been carefully written to provide an introduction to the environment, environmental science and environmental engineering. **Chapter 2**, Mass transfer is the movement of mass from one location to the other. It occurs through the processes of evaporation, absorption, drying, precipitation, distillation and membrane filtration. The exchange of energy, usually thermal energy between two or more physical systems is called heat or energy transfer. It occurs through the processes of conduction, convection or radiation. This chapter closely examines the different mechanisms of mass and energy transfer in the environment, such as evaporation, condensation, precipitation, infiltration, conduction, etc. **Chapter 3**, Ecology is a branch of biology. It studies the relationship among organisms and between organisms and the environment. To develop an understanding of the field, it is vital to delve into the fundamentals of plant and animal ecology, systems ecology and microbial ecology which have been extensively covered in this chapter. **Chapter 4**, Water is vital for all life on earth. It is a transparent, colorless and tasteless substance with the chemical formula H_2O. All the diverse aspects of water, the water cycle and water chemistry have been carefully examined in this chapter. **Chapter 5**, Natural hazards are natural phenomena, which have a detrimental effect on human beings and the environment. These can be broadly classified into geological and meteorological hazards. This chapter discusses in elaborate detail the different typess of natural hazards such as earthquake, coastal erosion, landslide, hurricane, tornado, flood, etc. **Chapter 6**, Hazardous wastes are the wastes that pose a potential threat to public health and safety. It can also cause harm to the environment. The topics elaborated in this chapter on household hazardous waste, radioactive waste and waste treatment technology, have been carefully written to provide a comprehensive understanding of hazardous wastes. **Chapter 7**, The introduction of any form of contaminant into the natural environment can cause harm to the environment and to life forms. This is termed as pollution. It can be of different forms. The topics covered in this chapter address the varied types of pollution such as air, water, noise and soil pollution and the different mitigation strategies for each. **Chapter 8**, The control and management of the human influence on the environment is under the scope of environmental resource management. It strives to ensure the protection and maintenance of ecosystem services for use by human societies in the future, while ensuring ecosystem integrity. This chapter discusses in extensive detail the fundamentals of water resource management, solid waste management, air resource management and environmental management.

At the end, I would like to thank all those who dedicated their time and efforts for the successful completion of this book. I also wish to convey my gratitude towards my friends and family who supported me at every step.

Carol Thorburn

Introduction to Environment Science and Engineering

Environmental science is an academic field that integrates biological, physical and information sciences for the study of the environment. A branch of engineering that is concerned with the protection of human populations from the effects of adverse environmental conditions is under the domain of environmental engineering. This chapter has been carefully written to provide an introduction to the environment, environmental science and environmental engineering.

Environment

The word Environment is derived from the French word "Environ" which means "surrounding". Our surrounding includes biotic factors like human beings, Plants, animals, microbes, etc. and abiotic factors such as light, air, water, soil, etc.

Environment is a complex of many variables, which surrounds man as well as the living organisms. Environment includes water, air and land and the interrelation ships which exist among and between water, air and land and human beings and other living creatures such as plants, animals and microorganisms environment consists of an inseparable whole system constituted by physical, chemical, biological, social and cultural elements, which are interlinked individually and collectively in myriad ways.

The natural environment consist of four interlinking systems namely, the atmosphere, the hydrosphere, the lithosphere and the biosphere. These four systems are in constant change and such changes are affected by human activities and vice versa.

Environmental Science

Environmental science is a field that deals with the study of interaction between human systems and natural systems. Natural systems involve the earth itself and life. Human systems are primarily the populations of the earth.

Environmental science is the academic field that takes physical, biological and chemical sciences to study the environment and discover solutions to environmental problems. Sciences used in environmental science include geography, zoology, physics, ecology, oceanology, and geology. Environmental science also branches out into environmental studies and environmental engineering. It provides integrated and interdisciplinary approach to the study of environmental problems.

Environmental studies are the study of social sciences to understand human interactions with the environment.

Environmental engineering is the focus on analyzing and deducing problems with the environment and the effect of man-made programs on the environment, and for finding solutions to help protect and preserve the environment by disposing of pollution in the air, water, and land.

Environmental science involves different fields of study. Most often, the study of environmental science includes the study of climate change, natural resources, energy, pollution, and environmental issues. In environmental sciences, ecologists study how plants and animals interact with each other, chemists study the living and non-living components of the environment, geologists study the formation, structure and history of earth, biologists study the biodiversity, Physicists are involved in thermodynamics, computer scientists are involved in technical innovations and computer modelling and biomedical experts study the impact of environmental issues on our health and social lives.

The growing complexity of environmental problems are creating a need for scientists with rigorous, interdisciplinary training in environmental science. Environmental scientists and specialists use their knowledge of the natural sciences to protect the environment and human health. They must have a solid background in economics, sociology and political science.

Importance of Environmental Science

To Realize that Environmental Problems are Global

Environmental science lets you recognize that environmental problems such as climate change, global warming, ozone layer depletion, acid rains, and impacts on biodiversity and marine life are not just national problems, but global problems as well. So, concerted effort from across the world is needed to tackle these problems.

To Understand the Impacts of Development on Environment

It's well documented and quantified that development results in Industrial growth, urbanization, expansion of telecommunication and transport systems, hi-tech agriculture and expansion of housing. Environmental science seeks to teach the general population about the need for decentralization of industries to reduce congestion in urban areas. Decentralization means many people will move out of urban centers to reduce pollution resulting from overpopulation. The goal is to achieve all this sustainably without compromising the future generation's ability to satisfy their own needs.

To Discover Sustainable Ways of Living

Environmental science is more concerned with discovering ways to live more sustainably. This means utilizing present resources in a manner that conserves their supplies for the future. Environmental sustainability doesn't have to outlaw living luxuriously, but it advocates for creating awareness about consumption of resources and minimizing unnecessary waste. This includes minimizing household energy consumption, using disposals to dispose of waste, eating locally, recycling more, growing your own food, drinking from the tap, conserving household water, and driving your car less.

To Utilize Natural Resources Efficiently

Natural resources bring a whole lot of benefits to a country. A country's natural resources may not be utilized efficiently because of low-level training and lack of management skills. Environmental science teaches us to use natural resources efficiently by:

- Appropriately putting into practice environmental conservation methods.

- Using the right tools to explore resources.

- Adding value to our resources.

- Making sure machines are maintained appropriately.

- Thorough training of human resources.

- Provision of effective and efficient supervision.

- Using the right techniques to minimize exploitation.

- To understand behavior of organisms under natural conditions.

Behavior is what organisms manifest to respond to, interact with, and control their environment. An animal exhibits behavior as the first line of defense in response to any change of environment. So, critical look at organism's behavior can offer insightful information about animal's needs, dislikes, preferences and internal condition providing that your evaluation of those observations firmly hinge on knowledge of species'-natural behavior.

Biodiversity is the variety of life on earth. The present rate of biodiversity loss is at an all-time high. Environmental science aims to teach people how to reverse this trend by:

- Using sustainable wood products.

- Using organic foods.

- Embracing the 3R's, reduce, reuse, and recycle.

- Purchasing sustainable seafood.

- Supporting conservation campaigns at local levels.

- Conserving power

- Minimizing consumption of meat

- Utilizing eco-friendly cleaning products

- To understand the interrelationship between organisms in population and communities

Organisms and humans depend on each other to get by. Environmental science is important because it enables you to understand how these relationships work. For example, humans breathe out carbon dioxide, which plants need for photosynthesis. Plants, on the other hand, produce and release oxygen to the atmosphere, which humans need for respiration. Animal droppings are sources of nutrients for plants and other microorganisms. Plants are sources of food for humans and animals. In short, organisms and humans depend on each other for survival.

Environmental problems at local, national and international levels mostly occur due to lack of awareness. Environmental science aims to educate and equip learners with necessary environmental skills to pass to the community in order to create awareness. Environmental awareness can be created through social media, creating a blog dedicated to creating awareness, community centered green clubs, women forums, and religious podiums.

Components of Environmental Science

Ecology

Ecology is the study of organisms and the environment interacting with one another. Ecologists, who make up a part of environmental scientists, try to find relations between the status of the environment and the population of a particular species within that environment, and if there is any correlations to be drawn between the two. For example, ecologists might take the populations of a particular type of bird with the status of the part of the Amazon Rainforest that population is living in.

The ecologists will study and may or may not come to the conclusion that the bird population is increasing or decreasing as a result of air pollution in the rainforest. They may also take multiple species of birds and see if they can find any relation to one another, allowing the scientists to come to a conclusion if the habitat is suitable or not for that species to live in.

Geoscience

Geoscience concerns the study of geology, soil science, volcanoes, and the Earth's crust as they relate to the environment. As an example, scientists may study the erosion of the Earth's surface in a particular area. Soil scientists, physicists, biologists, and geomorphologists would all take part in the study.

Geomorphologists would study the movement of solid particles (sediments), biologists would study the impacts of the study to the plants and animals of the immediate environment, physicists would study the light transmission changes in the water causing the erosion, and the soil scientists would make the final calculations on the flow of the water when it infiltrates the soil to full capacity causing the erosion in the first place.

Atmospheric Science

Atmospheric science is the study of the Earth's atmosphere. It analyzes the relation of the Earth's atmosphere to the atmospheres of other systems. This encompasses a wide variety of scientific studies relating to space, astrology and the Earth's atmosphere: meteorology, pollution, gas emissions, and airborne contaminants.

An example of atmospheric science is where physicists study atmospheric circulation of a part of the atmosphere, chemists would study the chemicals existent in this part and their relationships with the environment, meteorologists study the dynamics of the atmosphere, and biologists study how the plants and animals are affected and their relationship with the environment.

Environmental Chemistry

Environmental Chemistry is the study of the changes chemicals make in the environment, such as contamination of the soil, pollution of the water, degradation of chemicals, and the transport of chemicals upon the plants and animals of the immediate environment. An example of environmental chemistry would be introduction of a chemical object into an environment, in which chemists would then study the chemical bonding to the soil or sand of the environment. Biologists would then study the now chemically induced soil to see its relationship with the plants and animals of the environment.

Environmental science is an active and growing part of the scientific world accelerated by the need to address problems with the Earth's environment. It encompasses multiple scientific fields and sciences to see how all interchange and relate with one another in any of the above four components.

Environmental Engineering

Environmental engineering is the application of science and engineering principles to protect and enhance the quality of the environment—air, water, and land resources—to sustain the health of humans and other living organisms. Environmental engineers work on projects to conserve the environment, reduce waste, and cleanup sites that are already polluted. In so doing, they have to deal with a variety of pollutants—chemical, biological, thermal, radioactive, and even mechanical. In addition, they may become involved with public education and government policy-setting.

To meet its goals, environmental engineering incorporates elements from a wide range of disciplines, including chemistry, biology, ecology, geology, civil engineering, chemical engineering, mechanical engineering, and public health. Some consider environmental engineering to include the development of sustainable processes.

Scope of Environmental Engineering

There are several divisions in the field of environmental engineering.

Environmental Impact Assessment and Mitigation

This division is a decision-making tool. Engineers and scientists assess the impacts of a proposed project on environmental conditions. They apply scientific and engineering principles to evaluate the project's impacts on:

- The quality of air, water, habitat.
- Flora and fauna.
- Agricultural capacity.
- Traffic.
- Social needs and customs.

They also consider such factors as noise levels and visual (landscape) impacts.

If adverse impacts are expected, they then develop measures to limit or prevent such impacts. For example, to mitigate the filling-in of a section of wetlands during a proposed road development, they may plan for the creation of wetlands in a nearby location.

Water Supply and Treatment

Engineers and scientists work to secure water supplies for potable and agricultural use. They examine a watershed area and evaluate the water balance in terms of such factors as the availability of water for various needs and the seasonal cycles of water in the watershed. In addition, they develop systems to store, treat, and convey water for various uses. For example, for potable water supplies, water is treated to minimize the risk of diseases and to create a palatable water flavor. Water distribution systems are designed and built to provide adequate water pressure and flow rates to meet various needs, such as domestic use, fire suppression, and irrigation.

Wastewater Conveyance and Eatment

Most urban and many rural areas no longer discharge human waste directly to the land through outhouse, septic, or honey bucket systems. Rather, such waste is deposited into water and conveyed from households via sewer systems. Engineers and scientists develop systems to carry this waste material away from residential areas and to process it in sewage treatment facilities. In developed countries, substantial resources are applied to the treatment and detoxification of this waste before it is discharged into a river, lake, or ocean system. Developing nations are likewise striving to develop such systems, to improve water quality in their surface waters and reduce the risk of waterborne diseases.

There are numerous wastewater treatment technologies. A wastewater treatment train can consist of several systems:

1. A primary clarifier system to remove solid and floating materials.

2. A secondary treatment system, consisting of an aeration basin followed by flocculation and sedimentation, or an activated sludge system and a secondary clarifier. This system

removes organic material by growing bacteria (activated sludge). The secondary clarifier removes activated sludge from the water.

3. A tertiary biological nitrogen removal system and a final disinfection process. This system, although not always included due to costs, is becoming more prevalent. Its purpose is to remove nitrogen and phosphorus and to disinfect the water before discharge to a surface water stream or ocean outfall.

Air Quality Management

Engineers design manufacturing and combustion processes to reduce air emissions to acceptable levels. For example, devices known as scrubbers, precipitators, and after-burners are utilized to remove particulates, nitrogen oxides, sulfur oxides, and reactive organic gases from vapors, preventing their emission into the atmosphere. This area of work is beginning to overlap with the drive toward energy efficiency and the desire to reduce carbon dioxide and other greenhouse gas emissions from combustion processes. Scientists develop atmospheric dispersion models to evaluate the concentration of a pollutant at a source, or the impact on air quality and smog production from vehicle and flue-gas stack emissions.

Hazardous Waste Management

Hazardous waste is defined as waste that poses substantial or potential threats to public health or the environment, generally exhibiting one or more of the following characteristics: ignitability, corrosivity, reactivity, and toxicity. Hazardous wastes include:

- Industrial wastes, such as caustic and toxic chemicals used in manufacturing processes.

- Agricultural wastes, such as pesticides, herbicides, and excess nitrates and phosphates from fertilizers.

- Household wastes, such as paints, flammable solvents, caustic cleaners, batteries, pesticides, drugs, and mercury (from broken thermometers).

- Medical wastes, such as needles, scalpels, glassware, unused drugs, radioactive isotopes, and chemical wastes.

- Wastes from illegal drug manufacture, such as various harmful chemicals.

Hazardous wastes are commonly segregated into solid and liquid wastes. Solid hazardous wastes are generally taken to special landfills that are similar to conventional landfills but involving greater precautions to protect groundwater and workers. Liquid hazardous materials require highly specialized liners and treatment for disposal. These wastes are often stored in large outdoor manmade ponds and require extensive monitoring to protect groundwater and safeguard area residents.

Brownfield Land Management and Site Remediation

Brownfield lands, or simply "brownfields," are abandoned, idled, or under-used industrial and commercial sites where expansion or redevelopment is complicated by contamination with low levels of hazardous waste or other pollutants. These sites have the potential to be reused once they are cleaned up. Land that is severely contaminated, such as "Superfund" sites in the United States, does not fall under the brownfield classification.

Many contaminated brownfield sites sit idle and unused for decades, because of the cost of cleaning them to safe standards. The redevelopment of brownfield sites has become more common in the first decade of the twenty–first century, as developable land grows less available in highly populated areas, the methods of studying contaminated land become more precise, and techniques used to clean up environmentally distressed properties become more sophisticated and established.

Innovative remedial techniques employed at distressed brownfield properties include:

- Bioremediation: a remedial strategy that uses naturally occurring microbes in soils and groundwater to expedite cleanup.

- In-situ oxidation: a strategy that uses oxygen or oxidizing chemicals to enhance a cleanup.

- Soil vapor extraction: a process in which vapor from the soil phase is extracted and treated, thereby removing contaminants from the soil and groundwater beneath a site.

- Phytoremediation: an approach that utilizes deep-rooted plants to soak up metals in soils. When the plants reach maturity, they are removed and disposed of as hazardous wastes, as the metal contaminants have become part of the plants.

Often, these strategies are used in conjunction with one another, and the brownfield site is prepared for redevelopment.

Additional Applications

- Risk assessment

- Environmental policy and regulation development

- Environmental health and safety

- Natural resource management

- Noise pollution

Geographic Information System

The Geographic Information System (GIS) is a useful tool for environmental engineers as well as others. It consists of a computer system for collecting, storing, editing, analyzing, sharing, and displaying geographically-referenced information. GIS technology can be used for many applications, including environmental impact assessment, development planning, and resource management. For example, a GIS might be used to find wetlands that need protection from pollution.

2

Mass and Energy Transfer in Environment

Mass transfer is the movement of mass from one location to the other. It occurs through the processes of evaporation, absorption, drying, precipitation, distillation and membrane filtration. The exchange of energy, usually thermal energy between two or more physical systems is called heat or energy transfer. It occurs through the processes of conduction, convection or radiation. This chapter closely examines the different mechanisms of mass and energy transfer in the environment, such as evaporation, condensation, precipitation, infiltration, conduction, etc.

Evaporation

Evaporation is the process by which water (and other liquids) changes from a liquid state to a vapor or gas state.

Evaporation is great for separating a mixture (solution) of a soluble solid and a solvent.

The process involves heating the solution until the solvent evaporates (turns into gas) leaving behind the solid residue.

Here is a simple example involving a mixture of salt and water.

To get the salt back from the salt water, the solution is heated to boiling point. As it boils, the water escapes as vapor (gas). After some time, all the water evaporates, leaving a layer of salt at the bottom of the beaker.

Evaporation is also known as vaporization. Vaporization usually happens on the surface of the water. It is a slow process and occurs at all temperatures. For example, wet clothes take a lot of time to dry due to the slow rate of vaporization. Water bodies with larger surface tend to evaporate quicker.

For example, water spilt on a small part of the floor takes a long time to evaporate whereas, water spilt on the entire floor dries up rapidly. The vaporization rate also depends on the type of the liquid, its density, etc. Liquids like spirits, petrol, and kerosene evaporate quicker than water. Vaporization also depends on factors like temperature, humidity, pressure, wind speed, and surface area.

Particles take in energy from the surroundings during the process of vaporization. This absorption makes the surroundings cooler than before. For example, if you put acetone on your palm, you will feel a cooling sensation on your palm. Heat is absorbed from your palm by the particles, thus, your palm feels cool.

It has also been noticed that the temperature of the fresh tap water is higher than water kept in an open vessel. Only molecules with high kinetic energy can escape during the process of vaporization.

Factors That Affect Vaporization

The main factors that have an effect on vaporization are as follows:

- Vaporization increases with an increase in the surface area.

- It increases with an increase in temperature.

- Vaporization increases with the decrease in humidity.

- It increases with an increase in wind speed.

Evaporation Increases with an Increase in the Surface Area

If the surface area is increased, then the amount is of liquid that is exposed to air is larger. More molecules can escape with a wider surface area. For e.g. we spread out clothes to dry. We do that because that speeds up the process of vaporization.

Evaporation Increases with an Increase in Temperature

The water molecules move rapidly when the water is heated. This makes the molecules escape faster. Higher temperatures lead to increase in vaporization as more molecules get kinetic energy to convert into vapor. For example, boiling water evaporates faster than fresh tap water.

Evaporation Increases with a Decrease in Humidity

Humidity means the amount of vapor present in the air. The air around can only holds a certain amount of vapor at a certain time and certain temperature. If the temperature increases and the wind speed and humidity stay constant, then the rate of evaporation will increase since warmer air can hold more water vapor than colder air.

Evaporation increases with an increase in the speed of wind

Particles of vapor move away when the speed of wind increases. This leads to a decrease in the amount of water vapor in the atmosphere. For example, we use hand dryers to dry our hands. Here the wind is expelled from the hand dryer which dries our hand.

Condensation

Condensation is the change in phase of a substance from a gas (or vapor) to a liquid. It occurs when a vapor is cooled or compressed or subjected to both cooling and compression. Liquid that is formed by condensation of a vapor is called the condensate.

The phenomenon of condensation occurs regularly in nature. For example, clouds are formed by the condensation of water vapor in the atmosphere. Thus, condensation is an important process in the hydrologic cycle (water cycle). Also, the water seen on the outside of a cold glass on a hot day is condensation. In fundamental research and industrial processes, condensation is used in conjunction with distillation, as a means of separating a substance from a mixture.

Condensation of Water in Nature

When water vapor from the air naturally condenses on cold surfaces into liquid water, it is called dew. Condensation of water vapor within the atmosphere produces clouds.

Water vapor will only condense onto another surface when that surface is cooler than the temperature of the water vapor, or when the saturation humidity of the vapor in the air has been exceeded. When water vapor condenses onto a surface, the water releases some heat onto the surface, which becomes slightly warmer. At the same time, the temperature of the atmosphere drops very slightly.

The "dew point" (or "dew point temperature") is the temperature to which air must be cooled for it to become saturated with water vapor. Further cooling leads to condensation, with the formation of dew or fog. A net condensation of water vapor occurs on a surface when the temperature of the surface is at or below the dew point of the atmosphere. Deposition, the direct formation of ice from water vapor, is a type of condensation. Frost and snow are examples of deposition.

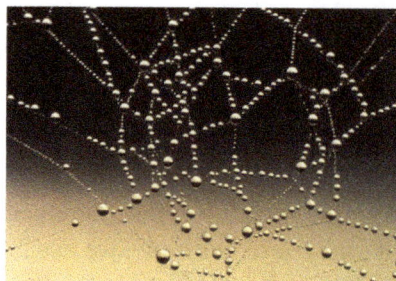

Condensation in Buildings

Condensation is the most common form of dampness encountered in buildings. In buildings, the internal air can have a high level of relative humidity due to the occupants' activities such as cooking, drying clothes, and even breathing. When this air comes into contact with cold surfaces such as windows and cold walls it can condense, causing dampness.

Applications of Condensation

Condensation is an important component of distillation, which is a means of separating a substance from a mixture, both in the research laboratory and industrial settings. A device used to condense vapors is called a condenser. Condensers are used in heat exchangers of various designs and come in many sizes, ranging from small (hand-held) units to very large systems.

Because condensation is a naturally occurring phenomenon, it can often be used to generate large quantities of water for human use. Many structures—such as fog fences, air wells, and dew ponds—are made solely for the purpose of collecting water from condensation. Such systems can be used to retain soil moisture in areas where active desertification is occurring. Some organizations provide education programs about water condensers in efforts to effectively aid such areas.

Precipitation

Precipitation is the falling of water from the sky in different forms. They all form from the clouds which are raised about 8 to 16 kilometers (4 to 11 miles) above the ground in the earth's troposphere. Precipitation takes place whenever any or all forms of water particles fall from these high levels of the atmosphere and reach the earth surface. The drop to the ground is caused by frictional drag and gravity. When one falling particle drops from the cloud, it leaves behind a turbulent wake, causing faster and continued drops.

The (clouds) crystallized ice may reach the ground as ice pellets or snow or may melt and change into raindrops before reaching the surface of the earth depending on the atmospheric temperatures. For this reason, there are many different types of precipitation namely rain, snow, sleet, freezing rain, hail, snow grains and diamond dust. They are forms of water that fall from the sky's frozen clouds.

Different Types of Precipitation

Rain

Rain is any liquid that drops from the clouds in the sky. Rain is described as water droplets of 0.5 mm or larger. Droplets less than half a millimeter are defined as drizzle. Raindrops frequently fall when small cloud particles strike and bind together, creating bigger drops. As this process continues, the drops get bigger and bigger to an extent where they become too heavy suspend on the air. As a result, the gravity pulls then down to the earth.

When high in the air, the raindrops start falling as ice crystals or snow but melt when as they proceed down the earth through the warmer air. Rainfall rates vary from time to time, for example, light rain ranges from rates of 0.01 to 0.1 inches per hour, moderate rain from 0.1 to .3 inches per hour, and heavy rain above 0.3 inches per hour. Rain is the most common component of the water cycle and replenishes most of the fresh water on the earth.

Snow

Snow occurs almost every time there is rain. However, snow often melts before it reaches the earth surface. It is precipitation in the form of virga or flakes of ice water falling from the clouds. Snow is normally seen together with high, thin and weak cirrus clouds. Snow can at times fall when the atmospheric temperatures are above freezing, but it mostly occur in sub-freezing air. When the temperatures are above freezing, the snowflakes can partially melt but because of relatively warm temperatures, the evaporation of the particles occurs almost immediately.

This evaporation leads to cooling just around the snowflake and makes it to reach to the ground as snow. Snow has fluffy, white and soft structure and its formation is in different shapes and ways, namely flat plates and thin needles. Each type of snow forms under specific combinations of atmospheric humidity and temperatures. The process of snow precipitation is called snowfall.

Sleet (Ice Pellets)

Sleet takes place in freezing atmospheric conditions. Sleet, also known as ice pellets, form when snow falls into a warm layer then melts into rain and then the rain droplets falls into a freezing layer of air that is cold enough to refreeze the raindrops into ice pellets. Hence, sleet is defined as a form of precipitation composed of small and semitransparent balls of ice. They should not be confused with hailstones as they are smaller in size.

Sleet is often experienced during thunderstorms and is normally accompanied with frosty ice crystals that form white deposits and a mixture of semisolid rain and slushy snow. Ice pellets (sleet) bounce when they hit the ground or any other solid objects and falls with a hard striking sound. Sleet don not freeze into a solid mass except when it combines with freezing rain.

Freezing Rain

Freezing rain happens when rain falls during below freezing conditions/temperatures. This normally results in the solidification of rain droplets. The raindrops are super-cooled while passing through the sub-freezing layer in the atmosphere and freezes by the time it reaches the ground. During freezing rains, it is common to witness an even coating of ice on cars, streets, trees, and power lines. The resulting coating of ice is called glaze and it can build up to a thickness of several centimeters. Freezing rains pose a huge threat to normal operations of roadway transportation, aircrafts, and power lines.

Hail

Hailstones are big balls and irregular lumps of ice that fall from large thunderstorms. Hail is purely a solid precipitation. As opposed to sleets that can form in any weather when there are thunderstorms, hailstones are predominately experienced in the winter or cold weather. Hailstones are mostly made up of water ice and measure between 0.2 inches (5 millimeters) and 6 inches (15 centimeters) in diameter. This ranges in size of a pea's diameter to that larger than a grapefruit.

For this reason, they are highly damaging to crops, tearing leaves apart and reducing their value. Violent thunderstorms with very strong updrafts usually have the capability to hold ice against the gravitational pull, which brings about the hailstones when they eventually escape and fall to the ground. So, hailstones are formed from super-cooled droplets that slowly freeze and results in sheet of clear ice.

Drizzle

Drizzle is very light rain. It is stronger than mist but less than a shower. Mist is a thin fog with condensation near the ground. Fog is made up of ice crystals or cloud water droplets suspended in the air near or at the earth's surface. Drizzle droplets are smaller than 0.5 millimeters (0.02 inches) in diameter. They arise from low stratocumulus clouds. They sometimes evaporate even before reaching the ground due to their minute size. Drizzle can be persistent is cold atmospheric temperatures.

Sun Shower

Sun shower is a precipitation event that is registered when rain falls while the sun shines. It occurs when the winds bearing rain together with rain storms are blown several miles away, thus giving rise to raindrops into an area without clouds. Consequently, sun shower is formed when single rain cloud passes above the earth's surface and the sun's rays penetrate through the raindrops. Most of the time, it is accompanied with the appearance of a rainbow.

Snow Grains

Snow grains are as very small white and opaque grains of ice. Snow grains are fairly flat and have diameter generally less than 1mm. They are almost equivalent to the size of drizzle.

Diamond Dust

Diamond dust is extremely small ice crystals usually formed at low levels and at temperatures below -30 °C. Diamond dust got its name from the sparkling effect which is created when light reflects on the ice crystals in the air.

Infiltration

Anywhere in the world, a portion of the water that falls as rain and snow infiltrates into the subsurface soil and rock. How much infiltrates depends greatly on a number of factors. Infiltration of precipitation falling on the ice cap of Greenland might be very small, whereas, as this picture of a stream disappearing into a cave in southern Georgia, USA shows, a stream can act as a direct funnel right into groundwater.

Some water that infiltrates will remain in the shallow soil layer, where it will gradually move vertically and horizontally through the soil and subsurface material. Some of the water may infiltrate deeper, recharging groundwater aquifers. If the aquifers are porous enough to allow water to move freely through it, people can drill wells into the aquifer and use the water for their purposes. Water may travel long distances or remain in groundwater storage for long periods before returning to the surface or seeping into other water bodies, such as streams and the oceans.

Factors Affecting Infiltration

- Precipitation: The greatest factor controlling infiltration is the amount and characteristics (intensity, duration, etc.) of precipitation that falls as rain or snow. Precipitation that infiltrates into the ground often seeps into streambeds over an extended period of time, thus a stream will often continue to flow when it hasn't rained for a long time and where there is no direct runoff from recent precipitation.

- Base flow: To varying degrees, the water in streams have a sustained flow, even during periods of lack of rain. Much of this "base flow" in streams comes from groundwater seeping into the bed and banks of the stream.

- Soil characteristics: Some soils, such as clays, absorb less water at a slower rate than sandy soils. Soils absorbing less water result in more runoff overland into streams.

- Soil saturation: Like a wet sponge, soil already saturated from previous rainfall can't absorb much more thus more rainfall will become surface runoff.

- Land cover: Some land covers have a great impact on infiltration and rainfall runoff. Vegetation can slow the movement of runoff, allowing more time for it to seep into the ground. Impervious surfaces, such as parking lots, roads, and developments, act as a "fast lane" for rainfall - right into storm drains that drain directly into streams. Agriculture and the tillage of land also change the infiltration patterns of a landscape. Water that, in natural conditions, infiltrated directly into soil now runs off into streams.

- Slope of the land: Water falling on steeply-sloped land runs off more quickly and infiltrates less than water falling on flat land.

- Evapotranspiration: Some infiltration stays near the land surface, which is where plants put down their roots. Plants need this shallow groundwater to grow, and, by the process of evapotranspiration, water is moved back into the atmosphere.

Subsurface Water

Diagram showing how precipitation water seeps into the ground, through an unsaturated zone to saturate the saturated zone below the water table. As precipitation infiltrates into the subsurface soil, it generally forms an unsaturated zone and a saturated zone. In the unsaturated zone, the voids—that is, the spaces between grains of gravel, sand, silt, clay, and cracks within rocks—contain both air and water. Although a lot of water can be present in the unsaturated zone, this water cannot be pumped by wells because it is held too tightly by capillary forces. The upper part of the unsaturated zone is the soil-water zone. The soil zone is crisscrossed by roots, openings left by decayed roots and animal and worm burrows, which allow the precipitation to infiltrate into the soil zone. Water in the soil is used by plants in life functions and leaf transpiration, but it also can evaporate directly to the atmosphere. Below the unsaturated zone is a saturated zone where water completely fills the voids between rock and soil particles.

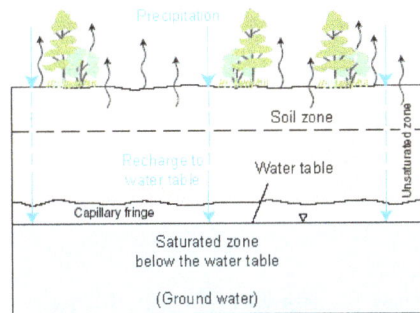

Natural refilling of deep aquifers is a slow process because groundwater moves slowly through the unsaturated zone and the aquifer. The rate of recharge is also an important consideration. It has been estimated, for example, that if the aquifer that underlies the High Plains of Texas and New Mexico—an area of slight precipitation—was emptied, it would take centuries to refill the aquifer at the present small rate of replenishment. In contrast, a shallow aquifer in an area of substantial precipitation such as those in the coastal plain in south Georgia, USA, may be replenished almost immediately.

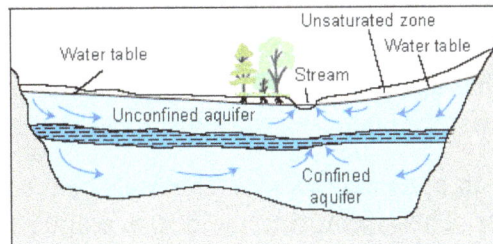

Artificial Recharge Gives Natural Infiltration a Push

People all over the world make great use of the water in underground aquifers all over the world. In fact, in some places, they pump water out of the aquifer faster than nature replenishes it. In these cases, the water table, below which the soil is saturated and possibly able to yield enough water that can be pumped to the surface, can be lowered by the excessive pumping. Wells can "go dry" and become useless.

In places where the water table is close to the land surface and where water can move through the aquifer at a high rate, aquifers can be replenished artificially. For example, large volumes of groundwater used for air conditioning are returned to aquifers through recharge wells on Long Island, New York.

Aquifers may be artificially recharged in two main ways:

- Rapid-infiltration pits: One way is to spread water over the land in pits, furrows, or ditches, or to erect small dams in stream channels to detain and deflect surface runoff, thereby allowing it to infiltrate to the aquifer.

- Groundwater injection: The other way is to construct recharge wells and inject water directly into an aquifer.

This picture shows rapid infiltration basins in Orlando, Florida. The water put into these basins recharges the shallow surficial aquifer and is used to irrigate local citrus crop fields.

Conduction

Conduction is one of the three main ways that heat energy moves from place to place. The other two ways heat moves around are radiation and convection. Conduction is the process by which heat energy is transmitted through collisions between neighboring atoms or molecules. Conduction occurs more readily in solids and liquids, where the particles are closer to together, than in gases, where particles are further apart. The rate of energy transfer by conduction is higher when there is a large temperature difference between the substances that are in contact.

Think of a frying pan set over an open camp stove. The fire's heat causes molecules in the pan to vibrate faster, making it hotter. These vibrating molecules collide with their neighboring molecules, making them also vibrate faster. As these molecules collide, thermal energy is transferred via conduction to the rest of the pan. If you've ever touched the metal handle of a hot pan without a potholder, you have first-hand experience with heat conduction.

Some solids, such as metals, are good heat conductors. Not surprisingly, many pots and pans have insulated handles. Air (a mixture of gases) and water are poor conductors of thermal energy. They are called insulators.

Conduction in the Atmosphere

Conduction, radiation and convection all play a role in moving heat between Earth's surface and the atmosphere. Since air is a poor conductor, most energy transfer by conduction occurs right near Earth's surface. Conduction directly affects air temperature only a few centimeters into the atmosphere.

During the day, sunlight heats the ground, which in turn heats the air directly above it via conduction. At night, the ground cools and the heat flows from the warmer air directly above to the cooler ground via conduction.

On clear, sunny days with little or no wind, air temperature can be much higher right near the ground that just a short way above. Although sunlight warms the surface, heat flow from the surface to the air above is limited by the poor conductivity of air. A series of thermometers mounted at different heights above the ground would reveal that air temperature falls off rapidly with height.

Convection

Convection is the flow of heat through a bulk, macroscopic movement of matter from a hot region to a cool region, as opposed to the microscopic transfer of heat between atoms involved with conduction. Suppose we consider heating up a local region of air. As this air heats, the molecules spread out, causing this region to become less dense than the surrounding, unheated air. For reasons discussed in the previous section, being less dense than the surrounding cooler air, the hot air will subsequently rise due to buoyant forces - this movement of hot air into a cooler region is then said to transfer heat by convection.

Heating a pot of water on a stove is a good example of the transfer of heat by convection. When the stove is first turned on heat is transferred first by conduction between the elements through the bottom of the pot to the water. However, eventually the water starts bubbling - these bubbles are actually local regions of hot water rising to the surface, thereby transferring heat from the hot water at the bottom to the cooler water at the top by convection. At the same time, the cooler, more dense water at the top will sink to the bottom, where it is subsequently heated. These convection currents are illustrated in the following figure.

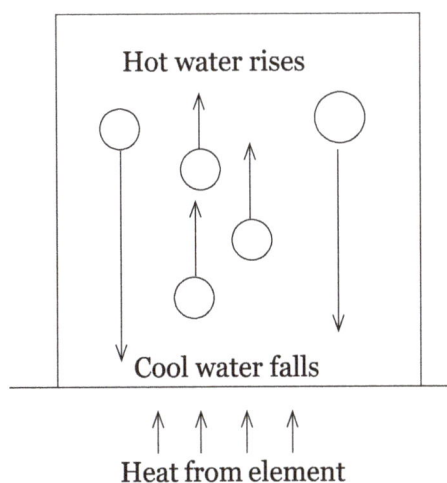

Consider now two regions separated by a barrier, one at a higher pressure relative to the other, and subsequently remove the barrier, as in the following figure.

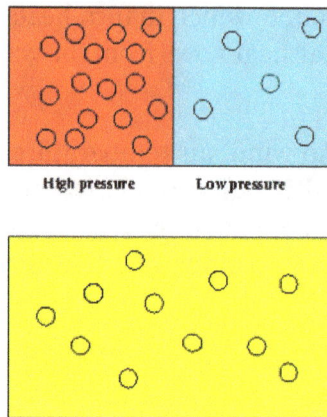

When the barrier is removed, material in the high pressure (high density) area will flow to the low pressure (low density) area. If the low pressure region was originally created by heating of the material, one sees that movement of material in this way is an example of heat flow by convection.

An important example of convection currents that can be interpreted in this manner is the creation of breezes over land masses next to large bodies of water. Water has a larger heat capacity than land, and subsequently holds heat better. It therefore takes longer to change its temperature, either upward or downward. Thus, during the day the air above the water will be cooler than that over the land. This creates a low pressure area over the land, relative to the high pressure area over the water, and subsequently one finds breezes blowing from the water to the land. On the other hand, during the night water cools off more slowly than the land, and the air above the water is slightly warmer than over the land. This creates a low pressure area over the water relative to the high pressure area over the land, and breezes will blow from the land to the water. These are illustrated in the following figure.

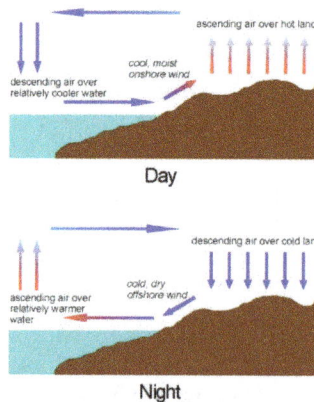

Radiation

Many forms of "radiation" are encountered in the natural environment and are produced by modern technology. Most of them have the potential for both beneficial and harmful effects. Even sunlight, the most essential radiation of all, can be harmful in excessive amounts. Most public attention is given to the category of radiation known as "ionizing radiation." This radiation can disrupt atoms, creating positive ions and negative electrons, and cause biological harm. Ionizing radiation includes x-rays, gamma rays, alpha particles, beta particles, neutrons, and the varieties of cosmic rays.

Radiation Damage and its Study

All ionizing radiations, at sufficiently large exposures, can cause cancer. Many, in carefully controlled exposures, are also used for cancer therapy. Whether harmful or beneficial, exposures to ionizing radiation have been an inevitable part of the environment throughout the Earth's history. The nucleosynthesis processes that produced the elements created both stable and unstable nuclides. The unstable nuclides with very long half-lives, together with their radioactive progeny, constitute the natural radioactivity on Earth today. In addition, violent processes in the sun and elsewhere lead to the bombardment of the Earth by cosmic rays. Thus, radiation is an old and familiar, if unrecognized, pollutant.

Information on the effects of radiation comes from studies of exposed groups and individuals, from animal experiments, and from studies at the cellular and molecular level. It is now well established that ionizing radiation has both prompt and delayed effects. At very high radiation exposures, death will occur within several months or less. At moderate levels, radiation exposure increases the chance that an individual will develop cancer, with a time delay of ten or more years for most cancers. At low levels, the cancer risk decreases, but the relationship between cancer risk and the magnitude of the exposure is uncertain.

Other effects of radiation, in part inferred from animal experiments, include an increased risk of genetic defects and, for exposures of the fetus before birth, of mental retardation. In terms of frequency of occurrence and severity of effects, cancer is the most serious consequence and receives the greatest attention.

Units of Radioactivity and Dose

The original unit for measuring the amount of radioactivity was the curie Ci first defined to correspond to one gram of radium-226 and more recently defined as:

$1 curie = 3.7 \times 10^{10}$ radioactive decays per second [exactly].

In the International System of Units (SI) the becquerel (Bq) has replaced the curie, where

$1 becquerel = 1$ radioactive decay per second $= 2.703 \times 10^{-11} Ci$

The magnitude of radiation exposures is specified in terms of the radiation dose.

There are two important categories of dose:

1. The *absorbed dose*, sometimes also known as the physical dose, defined by the amount of energy deposited in a unit mass in human tissue or other media. The original unit is the $[100\ erg\ /\ g]$; it is now being widely replaced by the SI unit, the gray (Gy) $[1\ J\ /\ kg]$, (Gy) [1 J/kg], where 1 gray = 100 rad.

2. The biological dose, sometimes also known as the dose equivalent, expressed in units of rem or, in the SI system, sievert (Sv). This dose reflects the fact that the biological damage caused by a particle depends not only on the total energy deposited but also on the rate of energy loss per unit distance traversed by the particle (or "linear energy transfer"). For example, alpha particles do much more damage per unit energy deposited than do electrons. This effect can be represented, in rough overall terms, by a quality factor, Q. Over a wide range of incident energies, Q is taken to be 1.0 for electrons (and for x-rays and gamma rays, both of which produce electrons) and 20 for alpha particles. For neutrons, the adopted quality factor varies from 5 to 20, depending on neutron energy.

The biological impact is specified by the dose equivalent H, which is the product of the absorbed dose D and the quality factor Q: $H = QD$.

The unit for the dose equivalent is the rem if the absorbed dose is in rads and the sievert (Sv) if the absorbed dose is in grays. Thus, $1\ Sv = 100\ rem$. As discussed below, 1 rem is roughly the average dose received in 3 years of exposure to natural radiation. 1 Sv is at the bottom of the range of doses that, if received over a short period of time, are likely to cause noticeable symptoms of radiation sickness.

The dose equivalent is still not the whole story. If only part of the body is irradiated, the dose must be discounted with an appropriate weighting factor if it is to reflect overall risk. The discounted dose is termed the effective dose equivalent or just the effective dose, expressed in rems or sieverts.

Radioactivity in the Natural Environment

The radioactive nuclei, or radionuclides, found naturally on Earth can be grouped into three series—headed by uranium-238, uranium-235, and thorium-232—plus several isolated beta-particle emitting nuclei, most prominently potassium-40 and rubidium-87. Average abundances of these nuclides are listed in table.

Half-lives and average abundances of natural radionuclides.

	$40\,\text{K}$	$87\,\text{Rb}$	$232\,\text{Th}$	$238\,\text{U}$
Half-life (billion years)	1.277	47.5	14.05	4.468
Upper continental crust				
Elemental abundance (ppm)	28000	112	10.7	2.8
Activity (Bq/kg)	870	102	43	35
Activity (nCi/kg)	23	2.7	1.2	0.9
Activity (kCi/km3)	66	8	3.3	2.6
Oceans				
Elemental concentration (mg/liter)	399	0.12	1×10^{-7}	0.0032
Activity (Bq/liter)	12	0.11	4×10^{-7}	0.040
Activity (nCi/liter)	0.33	0.003	1×10^{-8}	0.0011

Ocean sediments				
Elemental abundance (ppm)	17000		5.0	1.0
Activity (Bq/kg)	500		20	12
Activity (nCi/kg)	14		0.5	0.3
Human body				
Total activity (Bq)	4000	600	0.08	0.4a
Total activity (nCi)	100	16	0.002	0.01

a. In the human body the activity of ^{210}Pb and ^{210}Po, both progeny of ^{238}U, is much greater than that of ^{238}U itself.

The most interesting of the series is the uranium-238 series which decays via a chain containing 8 alpha decays and 6 beta decays to lead-206. This chain includes the longest-lived isotopes of radium and radon: radium-226 and radon-222, respectively. In each of the three chains the parent nucleus has a much greater lifetime than does any of the progeny. Therefore, a steady-state is established in which, for a given sample of material, each member of the series has the same activity—aside from deviations due to differences in chemical properties, which cause different elements to be transferred at different rates into or out of a given sample of material.

Including all the succeeding decays, the total activity in the thorium-232 and uranium-238 series is, very roughly, ten times the activity indicated for thorium-232 and uranium-238 alone. Thus, for each of the series, the total activity in the Earth's crust averages roughly $30,000\ Ci/km^3$. For both series together and including the contributions of potassium-40 and rubidium-87, the total activity in the crust averages about $100,000\ Ci/km^3$. There is also a considerable amount of radioactivity in the oceans, with potassium-40 dominant in the ocean itself and thorium-232 relatively more important in the ocean sediments. For the oceans as a whole $\left(1.4\times10^{21}\ liters\right)$ the total activity is about 4×10^{11} Ci for potassium $m-40\ and\ 1\times10^{9}\ Ci$ for uranium-238. Potassium-40 is also present in significant amounts in the human body, especially in muscle tissue.

In addition to these ancient radionuclides and their progeny, some radionuclides are being continually produced by cosmic rays. The most prominent of these is carbon-14, produced in the interaction of cosmic ray neutrons with nitrogen in the atmosphere.

Table: Average radiation doses in the United States, 1980-1982 (effective dose per year).

Radiation source	Comments	Effective dose	
		mSv/yr	mrem/y r
Natural sources	due to seepage of 222Rn from ground	2.0	200
indoor radon			
radionuclides vin	primarily 40K and 238U progeny	0.39	39
body			
terrestrial radiation	due to gamma-ray emitters in ground	0.28	28
cosmic rays	roughly doubles for 2000 m gain in elevation	0.27	27
cosmogenic	especially 14C	0.01	1
total (rounded)		3.0	300

Medical sources			
Diagnostic x-rays	excludes dental examinations	0.39	39
Medical treatments	radionuclides used in diagnosis (only)	0.14	14
total		0.53	53
Other			
consumer products	primarily drinking water, building materials	0.1	≈10
occupational	averaged over entire US population	0.01	1
nuclear fuel cycle	does not include potential reactor accidents	0.0005	0.05
TOTAL (rounded)		3.6	360

Effects of Low Doses

Most of the radiation doses that are received by members of the public and by radiation workers—both routinely and in accidents—are what are commonly referred to as "low doses." There is no precise definition of "low" but it would include doses below for example, 10 mSv per year. It is obviously important to determine the effects of low radiation doses—or, more precisely, the effects of small additions to the unavoidable natural background dose.

However, despite much study, these effects are not known, being too small to see unambiguously. The most prominent assumption, accepted by most official bodies, is the so-called linearity hypothesis, according to which the cancer risk is directly proportional to the magnitude of the dose, down to zero doses. In applying this assumption a consensus estimate is that the risk to a "typical" individual of an eventual fatal cancer is 0.00005 per mSv (or 0.05 per Sv). Thus, if 100,000 people each receive an added dose of 1 mSv, then 5 additional cancer deaths are to be expected. At the same time, while adopting the linearity hypothesis as a prudent working assumption, many of the leading studies have also indicated the possibility that small increases in radiation dose do not create any additional cancer risk. This reflects the considerable disagreement that exists within the scientific community as to the validity of the linearity hypothesis.

Effects of Large Doses

Radiation doses above 3 Gy (300 rad) can be fatal and doses above 6 Gy (600 rad) are almost certain to be fatal, with death occurring within several months (in shorter times at higher doses). Above 1 Gy, radiation causes a complex of symptoms, including nausea and blood changes, known as radiation sickness. For doses below 1 Sv (100 rem), there is little likelihood of radiation sickness, and the main danger is an increased cancer risk. The most important data base and analyses are from the RERF studies of the Hiroshima and Nagasaki survivors. In these studies, the exposure and medical histories are analyzed for an exposed group (50,113 people) and an unexposed, or minimally exposed, group (36,459 people). Through 1990, there have been 4,741 cancer fatalities in the exposed group, of which 454 are attributed to radiation exposure. There is a statistically significant excess for both solid cancer tumors and leukemia for doses above 0.2 Sv (20 rem). These data, in a succession of updated versions, have provided much of the information used in comprehensive studies of radiation effects.

Natural Background Sources

Natural background radiation comes from the following three sources:

Cosmic Radiation

The sun and stars send a constant stream of cosmic radiation to Earth, much like a steady drizzle of rain. Differences in elevation, atmospheric conditions, and the Earth's magnetic field can change the amount (or dose) of cosmic radiation that we receive.

Terrestrial Radiation

The Earth itself is a source of terrestrial radiation. Radioactive materials (including uranium, thorium, and radium) exist naturally in soil and rock. Essentially all air contains radon, which is responsible for most of the dose that Americans receive each year from natural background sources. In addition, water contains small amounts of dissolved uranium and thorium, and all organic matter (both plant and animal) contains radioactive carbon and potassium. Some of these materials are ingested with food and water, while others (such as radon) are inhaled. The dose from terrestrial sources varies in different parts of the world, but locations with higher soil concentrations of uranium and thorium generally have higher doses.

Radon gas contributes nearly two-thirds of our natural background radiation exposure.

Internal Radiation

All people have internal radiation, mainly from radioactive potassium-40 and carbon-14 inside their bodies from birth and, therefore, are sources of exposure to others. The variation in dose from one person to another is not as great as that associated with cosmic and terrestrial sources.

Manmade Sources

In general, the following man-made sources expose the public to radiation (the significant radioactive isotopes are indicated in parentheses):

- Medical sources (by far, the most significant man-made source)
- Diagnostic x-rays
- Nuclear medicine procedures (iodine-131, cesium-137, and others)
- Consumer products
- Building and road construction materials
- Combustible fuels, including gas and coal
- X-ray security systems
- Televisions
- Fluorescent lamp starters

- Smoke detectors (americium)

- Luminous watches (tritium)

- Lantern mantles (thorium)

- Tobacco (polonium-210)

- Ophthalmic glass used in eyeglasses

- Some ceramics

To a lesser degree, the public is also exposed to radiation from the nuclear fuel cycle, beginning with uranium mining and milling through the disposal of used (spent) fuel. In addition, the public receives minimal exposure from the transportation of radioactive materials and fallout from nuclear weapons testing and reactor accidents (such as Chernobyl).

References

- What-is-evaporation-method-of-separation, elements-mixtures-compounds, science: eschooltoday.com, Retrieved 11 April 2018

- Condensation: newworldencyclopedia.org, Retrieved 18 July 2018

- Different-types-of-precipitation, geography: eartheclipse.com, Retrieved 10 April 2018

- Environmental-radiation, radiation: ehs.iastate.edu, Retrieved 30 June 2018

- Evaporation-and-factors-affecting-it, matter-in-our-surroundings, chemistry: toppr.com, Retrieved 12 March 2018

3

Ecology

Ecology is a branch of biology. It studies the relationship among organisms and between organisms and the environment. To develop an understanding of the field, it is vital to delve into the fundamentals of plant and animal ecology, systems ecology and microbial ecology which have been extensively covered in this chapter.

Ecology or ecological science is the scientific study of the distribution and abundance of living organisms and how these properties are affected by interactions between the organisms and their environment. The environment of an organism includes both the physical properties, which can be described as the sum of local abiotic factors like climate and geology, as well as the other organisms that share its habitat.

Ecology may be more simply defined as the relationship between living organisms and their abiotic and biotic environment or as "the study of the structure and function of nature". In this later case, structure includes the distribution patterns and abundance of organisms, and function includes the interactions of populations, including competition, predation, symbiosis, and nutrient and energy cycles.

The term ecology (oekologie) was coined in 1866 by the German biologist Ernst Haeckel. The word is derived from the Greek oikos ("household," "home," or "place to live") and logos ("study")—therefore, "ecology" means the "study of the household of nature." The name is derived from the same root word as economics (management of the household), and thus ecology is sometimes considered the economics of nature, or, as expressed by Ernst Haeckel, "the body of knowledge concerning the economy of nature".

The interactions between living organisms and their abiotic and biotic environments, the focus of ecology, generally convey an overall sense of unity and harmony in nature. On the other hand, the history of the science itself has often revealed conflicts, schisms, and opposing camps, as ecologists took different approaches and often failed to meet on common ground.

Scope

Ecology is usually considered a branch of biology, the general science that studies living and once-living organisms. Organisms can be studied at many different levels, from proteins and nucleic acids (in biochemistry and molecular biology), to cells (in cellular biology), to multicellular systems (in physiology and anatomy, to individuals (in botany, zoology, and other similar disciplines), and finally at the level of populations, communities, and ecosystems, and to the biosphere as a whole. These latter strata, from populations to the biosphere, are the primary subjects of ecological inquiries.

Ecology is a multi-disciplinary science. Because of its focus on the higher levels of the organization of life on earth and on the interrelations between organisms and their environment, ecology draws heavily on many other branches of science, especially geology and geography, meteorology, pedology, chemistry, and physics. Thus, ecology is said to be a holistic science, one that overarches older disciplines, such as biology, which in this view become sub-disciplines contributing to ecological knowledge.

Agriculture, fisheries, forestry, medicine, and urban development are among human activities that would fall within Krebs' explanation of his definition of ecology: "where organisms are found, how many occur there, and why."

The term ecology is sometimes confused with the term environmentalism. Environmentalism is a social movement aimed at the goal of protecting natural resources or the environment, and which may involve political lobbying, activism, education, and so forth. Ecology is the science that studies living organisms and their interactions with the environment. As such, ecology involves scientific methodology and does not dictate what is "right" or "wrong." However, findings in ecology may be used to support or counter various goals, assertions, or actions of environmentalists.

Consider the ways an ecologist might approach studying the life of honeybees:

- The behavioral relationship between individuals of a species is behavioral ecology—for example, the study of the queen bee, and how she relates to the worker bees and the drones.

- The organized activity of a species is community ecology; for example, the activity of bees assures the pollination of flowering plants. Bee hives additionally produce honey, which is consumed by still other species, such as bears.

- The relationship between the environment and a species is environmental ecology—for example, the consequences of environmental change on bee activity. Bees may die out due to environmental changes. The environment simultaneously affects and is a consequence of this activity and is thus intertwined with the survival of the species.

Disciplines of Ecology

Ecology is a broad science which can be subdivided into major and minor sub-disciplines. The major sub-disciplines include:

- Physiological ecology (or Eco physiology), which studies the influence of the biotic and abiotic environment on the physiology of the individual, and the adaptation of the individual to its environment.

- Behavioral ecology, which studies the ecological and evolutionary basis for animal behavior, and the roles of behavior in enabling animals to adapt to their ecological niches.

- Population ecology (or autecology), which deals with the dynamics of populations within species and the interactions of these populations with environmental factors.

- Community ecology (or synecology) which studies the interactions between species within an ecological community.

- Ecosystem ecology, which studies the flows of energy and matter through ecosystems.

- Medical ecology, which studies issues of human health in which environmental disturbances play a role.

- Landscape ecology, which studies the interactions between discrete elements of a landscape and spatial patterns, including the role of disturbance and human impacts.

- Global ecology, which looks at ecological questions at the global level, often asking macro-ecological questions.

- Evolutionary ecology, which either can be considered the evolutionary histories of species and the interactions between them, or approaches the study of evolution by including elements of species interaction.

- And ecolinguistics, which looks at the relation between ecology and language.

Ecology can also be sub-divided on the basis of target groups:

- Animal ecology, plant ecology, insect ecology, human ecology, and so forth.

Ecology can, in addition, be sub-divided from the perspective of the studied biomes:

- Arctic ecology (or polar ecology), tropical ecology, desert ecology, aquatic ecology, terrestrial ecology, wetland ecology, and temperate zone ecology.

Ecology can also be sub-divided on whether or not the emphasis is on application to human activities, such as resource management, environmental conservation, and restoration:

- Theoretical ecology and applied ecology (including such subfields as landscape ecology, conservation biology, and restoration ecology).

Basic Concepts in Ecology

Ecology is a very broad-ranging and complex topic, and even its definition lacks consensus. Thus, there are numerous concepts that fit within this discipline, and diverse manners in which the content can be arranged and studied. Several of the basic concepts of ecology include ecological units, the ecosystem, energy flow, nutrient cycles, species interaction, productivity, and ecological challenges.

Ecological Units

For modern ecologists, ecology can be studied at several levels: population level (individuals of the same species), biocenosis level (or community of species), ecosystem level, biome level, and biosphere level.

The outer layer of the planet Earth can be divided into several compartments: the hydrosphere (or sphere of water), the lithosphere (or sphere of soils and rocks), and the atmosphere (or sphere of the air). The biosphere (or sphere of life), sometimes described as "the fourth envelope," is all living matter on the planet or that portion of the planet occupied by life. It reaches well into the other three spheres, although there are no permanent inhabitants of the atmosphere. Most life exists on or within a few meters of the Earth's surface. Relative to the volume of the Earth, the biosphere is only the very thin surface layer that extends from 11,000 meters below sea level to 15,000 meters above.

It is thought that life first developed in the hydrosphere, at shallow depths, in the photic zone (the area of water exposed to sufficient sunlight for photosynthesis). Multicellular organisms then appeared and colonized benthic zones. Terrestrial life developed later, after the ozone layer protecting living beings from UV rays formed. Diversification of terrestrial species is thought to be increased by the continents drifting apart, or alternately, colliding. Biodiversity is expressed at the ecological level (ecosystem), population level (intraspecific diversity), species level (specific diversity), and genetic level. Recently, technology has allowed the discovery of the deep ocean vent communities. This remarkable ecological system is not dependent on sunlight but bacteria, utilizing the chemistry of the hot volcanic vents, as the base of its food chain.

The biosphere contains great quantities of elements such as carbon, nitrogen, and oxygen. Other elements, such as phosphorus, calcium, and potassium, are also essential to life, yet are present in smaller amounts. At the ecosystem and biosphere levels, there is a continual recycling of all these elements, which alternate between their mineral and organic states.

A biome is a homogeneous ecological formation that exists over a vast region, such as tundra or steppes. The biosphere comprises all of the Earth's biomes—the entirety of places where life is possible—from the highest mountains to the depths of the oceans.

Biomes correspond rather well to subdivisions distributed along the latitudes, from the equator towards the poles, with differences based on the physical environment (for example, oceans or mountain ranges) and on the climate. Their variation is generally related to the distribution of species according to their ability to tolerate temperature and/or dryness. For example, one may find photosynthetic algae only in the photic part of the ocean (where light penetrates), while conifers are mostly found in mountains.

Though this is a simplification of a more complicated scheme, latitude and altitude approximate a good representation of the distribution of biodiversity within the biosphere. Very generally, biodiversity is greater near the equator and decreases as one approaches the poles.

The biosphere may also be divided into ecozones, which are biogeographical and ecological land classifications, such as Neartic, Neotropic, and Oceanic. Biozones are very well defined today and primarily follow the continental borders.

Ecological factors that can affect dynamic change in a population or species in a given ecology or environment is usually divided into two groups: biotic and abiotic.

Biotic factors relate to living organisms and their interactions. A biotic community is an assemblage of plant, animal, and other living organisms.

Abiotic factors are geological, geographical, hydrological, and climatological parameters. A biotope is an environmentally uniform region characterized by a particular set of abiotic ecological factors. Specific abiotic factors include:

- Water, which is at the same time an essential element to life.

- Air, which provides oxygen, nitrogen, and carbon dioxide to living species and allows the dissemination of pollen and spores.

- Soil, at the same time a source of nutriment and physical support (soil pH, salinity, nitrogen, and phosphorus content, ability to retain water and density are all influential).

- Temperature, which should not exceed certain extremes, even if tolerance to heat is significant for some species.

- Light, which provides energy to the ecosystem through photosynthesis.

- Natural disasters can also be considered abiotic.

The Ecosystem Concept

Some consider the ecosystem (abbreviation for "ecological system") to be the basic unit in ecology. An ecosystem is an ecological unit consisting of a biotic community together with its environment. Examples include a swamp, a meadow, and a river. It is generally considered smaller than a biome ("major life zone"), which is a large, geographic region of the earth's surface with distinctive plant and animal communities. A biome is often viewed as a grouping of many ecosystems sharing similar features, but is sometimes defined as an extensive ecosystem spread over a wide geographic area.

The first principle of ecology is that each living organism has an ongoing and continual relationship with every other element that makes up its environment. The ecosystem is composed of two entities, the entirety of life (the community, or biocoenosis) and the medium that life exists in (the biotope). Within the ecosystem, species are connected and dependent upon one another in the food chain, and exchange energy and matter between themselves and with their environment.

The concept of an ecosystem can apply to units of variable size, such as a pond, a field, or a piece of deadwood. A unit of smaller size is called a microecosystem. For example, an ecosystem can be a stone and all the life under it. A mesoecosystem could be a forest, and a macroecosystem a whole ecoregion, with its watershed.

Some of the main questions when studying an ecosystem include:

- How could the colonization of a barren area be carried out?

- What are the ecosystem's dynamics and changes?

- How does an ecosystem interact at local, regional, and global scale?

- Is the current state stable?

- What is the value of an ecosystem? How does the interaction of ecological systems provide benefit to humans, especially in the provision of healthy water?

Ecosystems are not isolated from each other, but are interrelated. For example, water may circulate between ecosystems by the means of a river or ocean current. Water itself, as a liquid medium, even defines ecosystems. Some species, such as salmon or freshwater eels move between marine systems and fresh-water systems. These relationships between the ecosystems lead to the concept of a biome.

Energy Flow

One focus of ecologists is to study the flow of energy, a major process linking the abiotic and biotic constituents of ecosystems.

While there is a slight input of geothermal energy, the bulk of the functioning of the ecosystem is based on the input of solar energy. Plants and photosynthetic microorganisms convert light into chemical energy by the process of photosynthesis, which creates glucose (a simple sugar) and releases free oxygen. Glucose thus becomes the secondary energy source that drives the ecosystem. Some of this glucose is used directly by other organisms for energy. Other sugar molecules can be converted to other molecules such as amino acids. Plants use some of this sugar, concentrated in nectar, to entice pollinators to aid them in reproduction.

Cellular respiration is the process by which organisms (like mammals) break the glucose back down into its constituents, water and carbon dioxide, thus regaining the stored energy the sun originally gave to the plants. The proportion of photosynthetic activity of plants and other photosynthesizes to the respiration of other organisms determines the specific composition of the Earth's atmosphere, particularly its oxygen level. Global air currents mix the atmosphere and maintain nearly the same balance of elements in areas of intense biological activity and areas of slight biological activity.

Nutrient Cycles

Ecologists also study the flow of nutrients in ecosystems. Whereas energy is not cycled, nutrients are cycled. Living organisms are composed mainly of carbon, oxygen, hydrogen, and nitrogen, and these four elements are cycled through the biotic communities and the geological world. These permanent recyclings of the elements are called biogeochemical cycles. Three fundamental bio-geochemical cycles are the nitrogen cycle, the water cycle, and the carbon-oxygen cycle. Another key cycle is the phosphorus cycle.

Water is also exchanged between the hydrosphere, lithosphere, atmosphere, and biosphere. The oceans are large tanks that store water; they ensure thermal and climatic stability, as well as the transport of chemical elements thanks to large oceanic currents.

Species Interactions

Biocenose, or community, is a group of populations of plants, animals, and microorganisms. Each population is the result of procreations between individuals of same species and cohabitation in a given place and for a given time. When a population consists of an insufficient number of individuals, that population is threatened with extinction; the extinction of a species can approach when all biocenoses composed of individuals of the species are in decline. In small populations, consanguinity (inbreeding) can result in reduced genetic diversity that can further weaken the biocenose.

Biotic ecological factors influence biocenose viability; these factors are considered as either intra-specific or interspecific relations.

Intraspecific relations are those which are established between individuals of the same species, forming a population. They are relations of cooperation or competition, with division of the territory, and sometimes organization in hierarchical societies.

Interspecific relations—interactions between different species—are numerous, and are usually de-scribed according to their beneficial, detrimental, or neutral effect (for example, mutualism or competition). Symbiosis refers to an interaction between two organisms living together in more or less intimate association. A significant relation is predation (to eat or to be eaten), which leads to the essential concepts in ecology of food chains (for example, the grass is consumed by the herbivore, itself consumed by a carnivore, itself consumed by a carnivore of larger size). A high predator-to-prey ratio can have a negative influence on both the predator and prey biocenoses in that low availability of food and high death rate prior to sexual maturity can decrease (or prevent the increase of) populations of each, respectively. Other interspecific relations include parasitism, infectious disease, and competition for limiting resources, which can occur when two species share the same ecological niche.

In an ecosystem, the connections between species are generally related to food and their role in the food chain. There are three categories of organisms:

- Producers: plants which are capable of photosynthesis.

- Consumers: animals, which can be primary consumers (herbivorous), or secondary or tertiary consumers (carnivorous).

- Decomposers: bacteria, mushrooms, which degrade organic matter of all categories, and restore minerals to the environment.

These relations form sequences in which each individual consumes the preceding one and is consumed by the one following, in what are called food chains or food networks.

The existing interactions between the various living beings go along with a permanent mixing of mineral and organic substances, absorbed by organisms for their growth, their maintenance, and their reproduction, to be finally rejected as waste. The interactions and biogeochemical cycles create a durable stability of the biosphere (at least when unchecked human influence and extreme weather or geological phenomena are left aside). This self-regulation, supported by negative feedback controls, supports the perenniality of the ecosystems. It is shown by the very stable concentrations of most elements of each compartment. This is referred to as homeostasis.

The ecosystem also tends to evolve to a state of ideal balance, reached after a succession of events, the climax (for example, a pond can become a peat bog).

Overall, the interactions of organisms convey a sense of unity and harmony. Plants, through photosynthesis, use carbon dioxide and provide oxygen, while animals use oxygen and give off carbon dioxide. On the level of the food web, plants capture the sun's energy and serve as food for herbivores, which serve as food for carnivores, and ultimately top carnivores. Decomposers (bacteria, fungi, etc.) break down organisms after they die into minerals that can be used by plants.

The harmony of species' interactions with other species and the environment, including the biogeochemical cycles, have proposed a theory by some that the entire planet acts as if one, giant, functioning organism. Lynn Margulis and Dorion Sagan in their book Microcosmos even propose that evolution is tied to cooperation and mutual dependence among organisms: "Life did not take over the globe by combat, but by networking."

The observed harmony can be attributed to the concept of dual purpose: the view that every entity in the universe in its interactions simultaneously exhibits purposes for the whole and for the individual—and that these purposes are interdependent. "Individual purpose" refers to the individual's requirement to met basic needs of self-preservation, self-strengthening, multiplication, and development. The "whole purpose" is that by which the individual contributes to the preservation, strengthening, and development of the larger entity of which it is a part. Thus, the cell of a multicellular body provides a useful function for the body of which it is part. This "whole purpose," which could be the secretion of an enzyme, harmonizes with the body's requirement of self-preservation, development, self-strengthening, and reproduction. The body, on the other hand, supports the cell's "individual purpose" by providing essential nutrients and carrying away wastes, assisting the cell's self-preservation, self-strengthening, multiplication, and development. Likewise, each individual organism exhibits both an individual purpose and a purpose for the whole related to its place in the environment. The result is an extraordinary harmony evident in creation.

Ecosystem Productivity

The concepts dealing with the movement of energy through an ecosystem (via producers, consumers, and decomposers) lead to the idea of biomass (the total living matter in a given place), of primary productivity (the increase in the mass of plants during a given time), and of secondary productivity (the living matter produced by consumers and the decomposers in a given time).

In any food network, the energy contained in the level of the producers is not completely transferred to the consumers. Thus, from an energy point of view, it is more efficient for humans to be primary consumers (to get nourishment from grains and vegetables) than as secondary consumers (from herbivores such as beef and veal), and more still than as tertiary consumers (from eating carnivores).

The productivity of ecosystems is sometimes estimated by comparing three types of land-based ecosystems and the total of aquatic ecosystems:

- The forests (one-third of the Earth's land area) contain dense biomasses and are very productive. The total production of the world's forests corresponds to half of the primary production.

- Savannas, meadows, and marshes (one-third of the Earth's land area) contain less dense biomasses, but are productive. These ecosystems represent the major part of what humans depend on for food.

- Extreme ecosystems in the areas with more extreme climates—deserts and semi-deserts, tundra, alpine meadows, and steppes—(one-third of the Earth's land area) have very sparse biomasses and low productivity.

- Finally, the marine and fresh water ecosystems (three-fourths of Earth's surface) contain very sparse biomasses (apart from the coastal zones).

Humanity's actions over the last few centuries have reduced the amount of the Earth covered by forests (deforestation), and have increased agro-ecosystems (agriculture). In recent decades, an increase in the areas occupied by extreme ecosystems has occurred (desertification).

Ecological Challenges

Generally, an ecological crisis is what occurs when the environment of a species or a population evolves in a way unfavorable to that species' survival.

It may be that environment quality degrades compared to the species needs, after a change in an abiotic ecological factor (for example, an increase of temperature, less significant rainfalls). It may be that the environment becomes unfavorable for the survival of a species (or a population) due to an increased pressure of predation (e.g., overfishing). It may be that the situation becomes unfavorable to the quality of life of the species (or the population) due to a rise in the number of individuals (overpopulation).

Although ecological crises are generally considered to be something that occurs in a short time span (days, weeks, or years), by definition, ecological crises can also be considered to occur over a very long time period, such as millions of years. They can also be of natural or anthropic origin. They may relate to one unique species or to many species.

Lastly, an ecological crisis may be local (an oil spill, a fire, or eutrophication of a lake), widespread (the movement of glaciers during an ice age), or global (a rise in the sea level).

According to its degree of endemism, a local crisis will have more or less significant consequences, from the death of many individuals to the total extinction of a species. Whatever its origin, disappearance of one or several species often will involve a rupture in the food chain, further impacting the survival of other species. Of course, what is an ecological crisis to one species, or one group of species, may be beneficial or neutral with respect to other species, at least short-term.

In the case of a global crisis, the consequences can be much more significant; some extinction events showed the disappearance of more than 90 percent of existing species at that time. However, it should be noted that the disappearance of certain species, such as the dinosaurs, by freeing an ecological niche, allowed the development and the diversification of the mammals. An ecological crisis may benefit other species, genera, families, orders, or phyla of organisms.

Sometimes, an ecological crisis can be a specific and reversible phenomenon at the ecosystem scale. But more generally, the crisis' impact will last. Indeed, it rather is a connected series of events that occur until a final point. From this stage, no return to the previous stable state is possible, and a new stable state will be set up gradually.

Lastly, if an ecological crisis can cause extinction, it can also more simply reduce the quality of life of the remaining individuals. Thus, even if the diversity of the human population is sometimes considered threatened, few people envision human disappearance at short span. However, epidemic diseases, famines, impact on health of reduction of air quality, food crises, reduction of living space, accumulation of toxic or non-degradable wastes, threats on key species (great apes, pandas, whales) are also factors influencing the well-being of people.

During the past decades, this increasing responsibility of humanity in some ecological crises has been clearly observed. Due to the increases in technology and a rapidly increasing population, humans have more influence on their own environment than any other ecosystem engineer.

Some examples as ecological crises are:

- Permian-Triassic extinction event—250 million of years ago.
- Cretaceous-Tertiary extinction event—65 million years ago.
- Ozone layer hole issue.
- Deforestation and desertification, with the disappearance of many species.

The nuclear meltdown at Chernobyl in 1986 that caused the death of many people and animals from cancer, and caused mutations in a large number of animals and people. The area around the plant is now abandoned because of the large amount of radiation generated by the meltdown.

Animal Ecology

Animal ecology concerns the relationships of individuals to their environments, including physical factors and other organisms, and the consequences of these relationships for evolution, population growth and regulation, interactions between species, the composition of biological communities, and energy flow and nutrient cycling through the ecosystem. From the standpoint of population, the individual organism is the fundamental unit of ecology. Factors influencing the survival and reproductive success of individuals form the basis for under-standing population processes.

Two general principles guide the study of animal ecology. One is the balance of nature, which states that ecological systems are regulated in approximately steady states. When a population becomes large, ecological pressures on population size, including food shortage, predation, and disease, tend to reduce the number of individuals. The second principle is that populations exist in dynamic relationship to their environments and that these relationships may cause ecological systems to vary dramatically over time and space. One of the challenges of animal ecology has been to reconcile these different viewpoints.

Populations depend on resources, including space, food, and opportunities to escape from predators. The amount of a resource potentially available to a population is generally thought of as being a property of the environment. As individuals consume resources they reduce the availability of these resources to others in the population. Thus, individuals are said to compete for resources. Larger populations result in a smaller share of resources per individual, which may lead to reduced survival and fecundity. Dense populations also attract predators and provide conditions for rapid transmission of contagious diseases, which generate pressure to reduce population size.

Changes in population size reflect both extrinsic variation in the environment that affects birth and death rates and intrinsic dynamics that result in oscillations or irregular fluctuations in population size. In some situations, the stable state may be a regular oscillation known as a limit cycle. Ecological systems also may switch between alternative stable states, as in the case of populations that are regulated at a high level by food limitation or at a low level by predators or other enemies. Switching between alternative stable states may be driven by changes in the environment.

Population Increase

In the absence of the effects of crowding, all populations have an immense capacity to increase. This capacity may be expressed as an exponential growth rate, which describes the growth of a population in terms of its relative, or percentage, rate of increase, like continuously compounded interest on a bank account. The constant r is often referred to as the Malthusian parameter. For a population growing at an exponential rate, the number of individuals (N) in a population at time t is $N(t) = (0)e^{rt}$ where N(0) is the number of individuals at time 0. Accordingly, the increase in a single time unit is er, which is the constant factor by which the population increases during each time period. The rate of increase in the number of individuals is then given by $dN/dt = rN$. The doubling time in years of a population growing exponentially is $t_2 = (ln\,2)/r$, or roughly $0.69/r$.

Estimated exponential annual growth rates of unrestrained populations range from low values of 0.077 for sheep in Tasmania and 0.091 for Northern elephant seals, to perhaps 1.0 for a pheasant population, 24 for the field vole, 10^{10} for flour beetles in laboratory cultures and 10^{30} for the water flea Daphnia. Human populations are at the lower end of this range, but a realistic exponential growth rate of 0.03 (or slightly above 3% per year) for some human populations is equivalent to a doubling time of about 23 years and a roughly thousand-fold increase in 230 years. Clearly, no population can maintain such a growth rate for long. (Expansion at the estimated annualized rates just cited for the field vole, flour beetle, and water flea is necessarily utterly fleeting.)

The exponential growth rate of a population can be calculated from the schedule of fecundity at age x (bx) and survival to age l) in a population. These "life table" variables are related to population growth rate by the Euler, or characteristic, equation,

Whose solution requires matrix methods? When the life table is unchanging for a long period, a population assumes a stable age distribution, which is also an intrinsic property of the life table, and a constant exponential rate of growth. Thus, assuming constant birth and death rates, the growth trajectory of a population may be projected into the future. However, because populations are finite and births and deaths are random events, the expected size of a population in the future has a statistical distribution that may include a finite probability of 0 individuals, that is, extinction. As a general rule, the probability of extinction decreases with increasing population size and increasing excess of births over deaths.

Population Regulation

Balancing the growth potential of all populations are various extrinsic environmental factors that act to slow population growth as the number of individuals increases. High population density depresses the resources of the environment, attracts predators, and, in some cases, results in stress-related reproductive failure or premature death. As population size increases, typically death rates of individuals increase, birth rates decrease, or both. The result is a slower growth rate and a changed, usually older, population. The predominant model used by animal ecologists to describe the relationship of population growth rate to population size (or density) is the logistic equation, in which the exponential growth rate of the population decreases linearly with increasing population size:

Where r_0 is the exponential growth rate of a population unrestrained by density (i.e., whose size is close to 0) and K represents the number of individuals that the environment can support at an equilibrium level, also referred to as the carrying capacity of the environment. Accordingly, the rate of growth of the population is expressed as Notice that when N < K, the growth rate is positive and the population grows. When N > K, the density-dependent term $(1 - N/K)$ is negative and the population declines. When N = K, the growth rate is 0 and a stable, steady-state population size is achieved. This depressing impact of density on the population growth rate is known as negative feedback.

The differential form of the logistic equation may be integrated to provide a function for the trajectory of population size over time.

The curve is sigmoid (S-shaped), with the rate of growth, dN/dt reaching a maximum (the inflection point) at N = K/2. Because this is the density at which individuals are added to the population most rapidly, the inflection point also represents the size of the population from which human consumers can remove individuals at the highest rate without causing the population to decline. Thus, the inflection point is also known as the point of maximum sustainable yield.

Density dependence can take on a variety of forms. One of these is a saturation model where the exponential growth rate remains constant and positive until a population completely utilizes a nonrenewable resource such as space, and population growth stops abruptly. The approach of a population to an equilibrium level determined by density-dependent processes can be altered by environmentally induced changes in the intrinsic rate of population growth or in the carrying capacity of the environment.

Difficulties in finding mates and maintaining other social interactions at low densities, including group defense against predators, may also cause the population growth rate to decrease as density declines (the Allee effect), and, below a certain density threshold, may even result in population decline to extinction. This type of response is a positive feed-back, one that promotes population instability. For example, after commercial hunting had reduced populations of the passenger pigeon to low levels, the decline in social interactions in this communally nesting species is thought to have doomed it to extinction.

Populations have inherent oscillatory properties that can be triggered by time lags in the response to changing density and which cause populations to fluctuate in a perpetual limit cycle, with alternating population highs and lows. In these cases higher values of r can send a population into unpredictable chaotic behavior, increasing the risk of extinction. In a population with continuous reproduction, regular population cycles occur when there is a lag, often equal to the period of development, in the response of a population to its own density effects on the environment. When the time lag is of period τ, limit cycles develop when $r\tau$ exceeds $\pi/2$, and the period of the cycle is 4 to 5 times τ.

Metapopulations

Most natural populations consist of many subpopulations occupying patches of suitable habitat surrounded by unsuitable environments. Oceanic islands and freshwater ponds are obvious examples. But fragmentation of forest and other natural habitats resulting from clearing land for agriculture or urban development is increasingly creating fragmented populations in many other kinds of habitats. These subpopulations are connected by movement of individuals, and the set of subpopulations is referred to as a metapopulation. Metapopulations have their own dynamics determined by the probabilities of colonization and extinction of individual patches. A set of simple metapopulation models describes changes in the proportion of patches occupied (). When the extinction probability (e) of an individual patch is independent of p, the rate of loss of subpopulations is simply pe. The rate of colonization is proportional to the number of patches that can provide potential colonists and the proportion of empty patches that are available to receive them. Hence, colonization is equal to cp(1-p), where c is the rate of colonization.

The metapopulation achieves a steady state of number of patches occupied when colonization balances extinction, that is $pe = cp(1-p)$, or $\hat{p} = 1 - e/c$. In this model, as long as the rate of colonization exceeds that of extinction, the metapopulation will persist. In more complex models, particularly when the probability of population extinction is reduced by continuing migration of individuals between patches (which keeps the sizes of subpopulations from dropping perilously low), the extinction rate and colonization rate both depend on patch occupancy. In this case, the solution to the metapopulation model has a critical ratio of colonization to extinction, below which patch occupancy declines until the metapopulation disappears. Thus, changes in patch size or migration between patches can cause an abrupt shift in the probability of metapopulation persistence.

Predator-prey Interactions

The dynamics of populations are influenced by interactions with predator and consumer populations. Because these interactions have built-in lag times in population responses, they often result in complex dynamics. Among the most spectacular fluctuations in size are those in populations of snowshoe hares and the lynx that prey on them. Population highs and lows may differ by a factor of 1,000 over an oscillation period of about ten years. Oscillation periods in other population cycles of mammals and birds in boreal forest and tundra habitats may be either approximately four years or nine to ten years.

The biologists Alfred Lotka and Vito Volterra independently developed models for the cyclic behavior of predator-prey systems in the 1920s. The most basic model expresses the rate of increase in the prey population in terms of the intrinsic growth capacity of the prey population and removal of prey individuals by predators, which is proportional to the product of the predator and prey population sizes. The growth of the predator population is equal to its birth rate, which depends on how many prey is captured, minus a density-independent term for the death of predator individuals. The joint equilibrium of the prey and predator populations is determined by the predation efficiency and the relative rates of birth and death of the prey and predator, respectively. However, the equilibrium is neutral, which means that any perturbation will set the system into a persisting cycle. More complex models of predator-prey interactions include a balance between various stabilizing factors, such as density-dependent control of either population, alternative food resources for predators, and refuges from predators at low prey densities, and destabilizing factors, such as time lags in the response of the predator and prey to each other. For the most part, these models predict stable predator and prey populations under constant conditions.

Both empirical and experimental studies have shown that the rate of predation is nonlinear, violating one of the assumptions of the Lotka-Volterra model. When predation is inefficient at low prey densities and predator populations are limited by density-dependence at high predator densities, there may be two stable points. One of these is at a high prey population level limited by the prey population's own food supply, the other at a low prey population level limited by predators. When a prey population, such as a crop pest, is released from predator control following depression of the predator population by extrinsic factors such as climate, disease, pesticides, and so on, the prey may increase to outbreak levels and become a severe problem. Thus, agricultural practices that incidentally depress the populations of natural control organisms can have unwanted consequences.

A special kind of predator-prey model is required to describe the interactions between parasites, including disease-causing organisms, and their hosts. These models need to take into account the fact that parasites generally do not kill their hosts, that the spread of parasites among hosts may depend on population density and the presence of suitable vectors, and that hosts may raise defensive immune reactions. Immune reactions create a time lag in the responses of parasite and host populations to each other and may result in strong fluctuations in the prevalence of parasitic diseases.

Plant Ecology

Plant ecology is a sub-discipline of ecology focused on the distribution and abundance of plants, and their interactions with the biotic and abiotic environment.

Plants are mostly sessile and photosynthetic organisms, and must attain their light, water, and nutrient resources directly from the immediate environment. Plant size and position in the community affect the capture and utilization of these resources and hence plants have evolved specific adaptations to enhance these capabilities. Understory plants have evolved mechanisms that allow them to tolerate low light conditions, while plants in the open have different mechanisms to cope with excess light. The absorption by roots and movement of water in the plant are determined by gradients in potential energy between the soil and atmosphere, as well as within the plant, as expressed by the concept of water potential. Nutrients are available through biological and chemical processes in the soil.

Mycorrhizae are critical in absorption of phosphorus and are also capable of interconnecting plants through their hyphae, thus facilitating belowground transfers of nutrients and water. Plants possess various adaptive functions, such as different photosynthetic pathways, that provide greater fitness in certain environments. In addition, there are correlations among plant traits, such as a positive relationship between photosynthetic rate and leaf nitrogen, or between leaf mass per area and photosynthesis, which suggest that there are ecological rules governing functional traits that cross species lines. Resource competition occurs when one or more resources are in limited supply and plants have various adaptations that maximize competitive success, including.

Allelopathy, when one plant releases an organic material into the environment to the detriment of a second plant. Plants also greatly influence the belowground environment (the rhizosphere) by altering the composition of the microbial community of bacteria and fungi. Interactions between above and belowground processes affect competitive outcomes and can alter community dynamics, including the process of successional change.

Primary succession occurs on new substrate and secondary succession occurs where vegetation previously existed. Secondary successions are initiated by disturbances such as fire, wind damage, flooding, grazing, and disease. Disturbance frequency and intensity greatly determine the development of the plant community and current and future climate change may result in new communities not present under present conditions, nor that resemble any from the recent past, making predictions of such impacts difficult.

Systems Ecology

Systems ecology is then the study of ecosystems that uses mathematical modeling, computation and, as the name implies, is based on systems theory. As with other areas of systems science, the use of systems theory as an approach involves the adoption of a holistic paradigm based on synthetic reasoning, meaning that systems ecology seeks a holistic view of the interactions and transactions within and between biological and geological systems on various scales. With this alternative approach, it does not restrict itself simply to the study of natural biophysical processes, but systems ecology now gives equal attention to the human dimension. Whereas standard ecology sees human industrial and economic activity as largely outside of its domain, systems ecologists recognize that the function of any ecosystem can be influenced by human economics in fundamental ways and that human industrial economic activity is a fundamental part of ecosystems around the world today. It has therefore taken an additional trans disciplinary step by including economics in the consideration of coupled socio-ecological systems. As such systems ecology takes an expansive domain of interest crossing almost all areas, from physics to biology, to economics and social studies to truly try and understand the workings of earth's systems in all their multi-dimensional complexity.

Microbial Ecology

Microbial ecology is the study of interrelationships between microorganisms and their living and nonliving environments. Microbial populations are able to tolerate and to grow under varying environmental conditions, including habitats with extreme environmental conditions such as hot springs and salt lakes. Understanding the environmental factors controlling microbial growth and survival offers insight into the distribution of microorganisms in nature, and many studies in microbial ecology are concerned with examining the adaptive features that permit particular microbial species to function in particular habitats.

Within habitats some microorganisms are autochthonous (indigenous), filling the functional niches of the ecosystem and others are allochthonous (foreign), surviving in the habitat for a period of time but not filling the ecological niches. Because of their diversity and wide distribution, microorganisms are extremely important in ecological processes. The dynamic interactions between microbial populations and their surroundings and the metabolic activities of microorganisms are essential for supporting productivity and maintaining environmental quality of ecosystems. Microorganisms are crucial for the environmental degradation of liquid and solid wastes and various pollutants and for maintaining the ecological balance of ecosystems—essential for preventing environmental problems such as acid mine drainage and eutrophication.

The various interactions among microbial populations and between microbes, plants, and animals provide stability within the biological community of a given habitat and ensure conservation of the available resources and ecological balance. Interactions between microbial populations can have positive or negative effects, either enhancing the ability of populations to survive or limiting population densities. Sometimes they result in the elimination of a population from a habitat.

The transfer of carbon and energy stored in organic compounds between the organisms in the community forms an integrated feeding structure called a food web. Microbial decomposition of dead plants and animals and partially digested organic matter in the decay portion of a food web is largely responsible for the conversion of organic matter to carbon dioxide.

Only a few bacterial species are capable of biological nitrogen fixation. In terrestrial habitats, the microbial fixation of atmospheric nitrogen is carried out by free-living bacteria, such as Azotobacter, and by bacteria living in symbiotic association with plants, such as Rhizobium or Bradyrhizobium living in mutualistic association within nodules on the roots of leguminous plants. In aquatic habitats, cyanobacteria, such as Anabaena and Nostoc, fix atmospheric nitrogen. The incorporation of the bacterial genes controlling nitrogen fixation into agricultural crops through genetic engineering may help improve yields. Microorganisms also carry out other processes essential for the biogeochemical cycling of nitrogen.

The biodegradation (microbial decomposition) of waste is a practical application of microbial metabolism for solving ecological problems. Solid wastes are decomposed by microorganisms in landfills and by composting. Liquid waste (sewage) treatment uses microbes to degrade organic matter, thereby reducing the biochemical oxygen demand (BOD).

References

- Ecology,entry: newworldencyclopedia.org: Retrieved 12 March 2018
- Plant-ecology, earth-and-planetary-sciences: sciencedirect.com, Retrieved 19 May 2018
- System-ecology: complexitylabs.io, Retrieved 20 June 2018
- Microbial-ecology: encyclopedia2.thefreedictionary.com, Retrieved 09 July 2018
- Animal-ecology, encyclopedias-almanacs-transcripts-and-maps, social-sciences: encyclopedia.com, Retrieved 31 March 2018

4

Water

Water is vital for all life on earth. It is a transparent, colorless and tasteless substance with the chemical formula H_2O. All the diverse aspects of water, the water cycle and water chemistry have been carefully examined in this chapter.

Water is a common chemical substance that is essential for all known forms of life. In typical usage, the term water refers to its liquid state, but the substance also has a solid state, ice, and a gaseous state, water vapor. About 71 percent of the Earth's surface is covered by water, mostly in oceans and other large water bodies.

The presence of water on Earth depends on various factors, including the Earth's location in the Solar System. If Earth were about 5 percent closer to or farther from the Sun, there would have been a much lower likelihood for the three forms of water to be present on this planet. Also, the Earth's mass is appropriate for gravity to hold an atmosphere, in which water vapor (along with carbon dioxide) helps maintain a relatively steady surface temperature. A smaller Earth would have a thinner atmosphere, causing temperature extremes and preventing the accumulation of water except at the polar ice caps. If Earth were much more massive, the water on it could have been in the solid state even at relatively high temperatures, because of the high pressure caused by gravity.

Water moves continually through a cycle of evaporation or transpiration, precipitation, and run-off, usually reaching the sea. Winds carry water vapor over land at the same rate as runoff into the sea, about 36 Tt per year. Over land, evaporation and transpiration contribute another 71 Tt per year to the precipitation of 107 Tt per year over land. Some water is trapped for varying periods in ice caps, glaciers, aquifers, or in lakes, sometimes providing freshwater for life on land. Water is a good solvent for a wide variety of substances.

Humans use water for many purposes, including drinking, cooking, cleaning, heating, and cooling. We find it valuable for scientific experimentation and industrial processes as well as for agriculture. In addition, we use water for various sports and recreational activities. In various religions, water is considered a purifier in an internal, spiritual sense as well as in an external, physical sense. Also, the Jordan River, Ganges River, and other bodies of water are considered sacred by people of certain religions.

Yet, water pollution, overconsumption, and uneven distribution have resulted in shortages of clean freshwater in many parts of the world. These shortages have in turn led to disputes between peoples of different nations.

Beyond the Earth, a significant quantity of water is thought to exist underground on the planet Mars, on Jupiter's moon Europa and Saturn's moon Enceladus, and on the exoplanets named HD 189733 b and HD 209458 b.

Water on Earth

Water is found in a variety of locations on Earth, in solid, liquid, and gaseous states. Accordingly, it is known by different names: water vapor and clouds in the sky; seawater and icebergs in the ocean; glaciers and rivers in the mountains; and aquifers in the ground. About 1,460 teratonnes (Tt) of water covers about 71 percent of the Earth's surface. Saltwater oceans hold 97 percent of surface water, glaciers and polar ice caps 2.4 percent, and other land surface water such as rivers and lakes 0.6 percent.

Origin and Planetary Effects

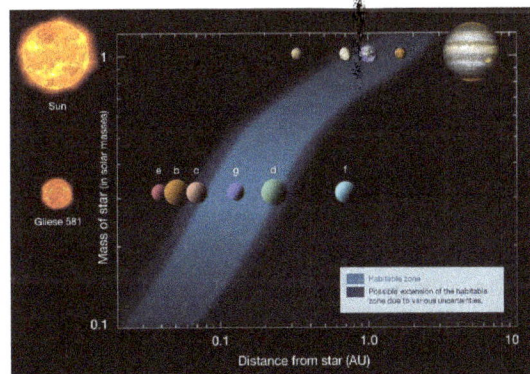

It is thought that much of the universe's water may have been produced as a by-product of star formation. The birth of a star is accompanied by a strong outward wind of gas and dust. When this outflow of material eventually impacts the surrounding gas, the resultant shock waves compress and heat the gas. Water could be quickly produced in this warm, dense gas.

Earth's Habitability

The existence of liquid water and to a lesser extent its gaseous and solid forms, on Earth is vital to the existence of life on Earth. The Earth is located in the habitable zone of the Solar System. If it were slightly closer to or farther from the Sun (about 5 percent, or 8 million kilometers or so), the conditions that allow the three forms of water to be present simultaneously would be far less likely to prevail.

Earth's mass allows its gravity to hold an atmosphere. Water vapor and carbon dioxide in the atmosphere provide a greenhouse effect that helps maintain a relatively steady surface temperature. If Earth were smaller, a thinner atmosphere would cause temperature extremes, preventing the accumulation of water except at the polar ice caps (as on Mars). If Earth were too massive, the water on it could have been in the solid state even at relatively high temperatures, because of the high pressure caused by gravity.

It has been proposed that life itself may maintain the conditions that have allowed its continued existence. Earth's surface temperature has been relatively constant through geologic time, despite varying levels of incoming solar radiation (insolation), indicating that a dynamic process governs Earth's temperature via a combination of greenhouse gases and surface or atmospheric albedo. This proposal is known as the Gaia hypothesis.

Tides

Tides are the cyclic rising and falling of Earth's ocean surface caused by the tidal forces of the Moon and the Sun acting on the oceans. Tides cause changes in the depth of the marine and estuarine water bodies and produce oscillating currents known as tidal streams. The changing tide produced at a given location is the result of the changing positions of the Moon and Sun relative to the Earth coupled with the effects of Earth rotation and the local bathymetry. The strip of seashore that is submerged at high tide and exposed at low tide, the intertidal zone, is an important ecological product of ocean tides.

Tastes and Odors of Water

Given that water can dissolve many different substances, it acquires different tastes and odors. In fact, humans and animals have developed senses to be able to evaluate the potability of water. Animals generally dislike the taste of salty sea water and the putrid swamps and favor the purer water of a mountain spring or aquifer. The taste advertised in spring water or mineral water derives from the minerals dissolved in it, as pure H_2O is tasteless. The "purity" of spring and mineral water refers to the absence of toxins, pollutants, and harmful microbes.

Effects on Life

Water has many distinct properties that are critical for the proliferation of all known forms of life, setting it apart from other substances. It is vital both as a solvent in which many of the body's solutes dissolve and as an essential part of many metabolic processes within the body, including reactions that lead to cellular replication and growth.

Metabolism is the sum total of anabolism and catabolism. In anabolism, water is removed from molecules (through energy-requiring enzymatic reactions) to build larger molecules (such as starches, triglycerides, and proteins for storage of fuels and information). In catabolism, water is used to break bonds, to generate smaller molecules (such as glucose, fatty acids, and amino acids). Water is thus essential and central to these metabolic processes. Without water, these metabolic processes would cease to exist.

Biochemical reactions take place in water at specific pH values. For instance, human enzymes usually perform optimally around a pH of 7.4. Digestion of food in the stomach requires the activity of an acid (hydrochloric acid, HCl). Some people suffer from what is called "acid reflux," in which the stomach acid makes its way into and adversely affects the esophagus. This condition can be temporarily neutralized by ingestion of a base such as aluminum hydroxide to produce the neutral molecules of water and aluminum chloride (a salt).

Water is also central to photosynthesis and respiration. Photosynthetic cells use the Sun's energy to split off water's hydrogen from oxygen. Hydrogen is combined with carbon dioxide (absorbed from air or water) to form glucose and release oxygen. All living cells use such fuels and oxidize the hydrogen and carbon to capture the Sun's energy and reform water and carbon dioxide in the process (cellular respiration).

Aquatic Life Forms

Earth's waters are filled with life. Nearly all fish live exclusively in water, and many marine mammals, such as dolphins and whales, also live in the water. Some kinds of animals, such as amphibians, spend portions of their lives in water and portions on land. Plants such as kelp and algae grow in the water and are the basis for some underwater ecosystems. Plankton is generally the foundation of the ocean food chain.

Different water creatures use different ways of obtaining oxygen in the water. Fish have gills instead of lungs, although some species of fish, such as the lungfish, have both. Marine mammals, such as dolphins, whales, otters, and seals, need to surface periodically to breathe air.

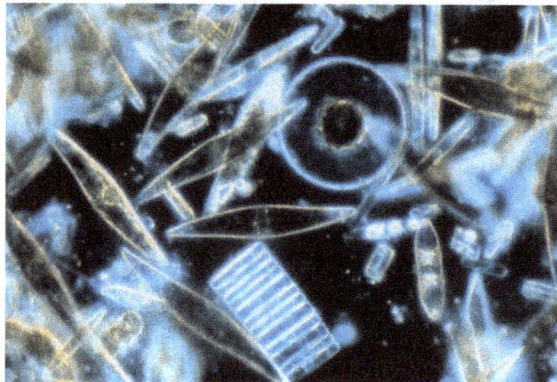

Human uses

Civilization has historically flourished around rivers and major waterways. Mesopotamia, the so-called cradle of civilization, was situated between the major rivers Tigris and Euphrates; the ancient Egyptians depended greatly upon the Nile. Large metropolitan areas like Rotterdam, London, Montreal, Paris, New York City, Shanghai, Tokyo, Chicago, Mumbai, and Hong Kong owe their success in part to their easy accessibility via water and the resultant expansion of trade. Islands with safe water ports, like Singapore, have flourished for the same reason. In regions such as North Africa and the Middle East, where freshwater is relatively scarce, access to clean drinking water has been a major factor in human development.

Water fit for human consumption is called drinking water or potable water. Water that is not potable can be made potable by various methods, including: filtration, to remove particulate impurities; chemical or heat treatment, to kill bacteria; and distillation, to separate water from impurities by vaporization and condensation. It should be noted, however, that some solutes in potable water are acceptable and even desirable for taste enhancement and to provide needed electrolytes.

Water that is not fit for drinking but is not harmful if used for swimming or bathing is sometimes called "safe water" or "safe for bathing." Chlorine, a skin and mucous membrane irritant, is used to make water safe for bathing or drinking. Its use is highly technical and is usually monitored by government regulations (typically 1 part per million (ppm) for drinking water, and 1-2 ppm of chlorine not yet reacted with impurities for bathing water).

Drinking Water

About 70 percent of the fat-free mass of the human body is made of water. To function properly, the body requires between one and seven liters of water per day to avoid dehydration; the precise amount depends on the level of activity, temperature, humidity, and other factors. Most of this is ingested through foods or beverages other than drinking straight water. It is not clear how much water intake is needed by healthy people.

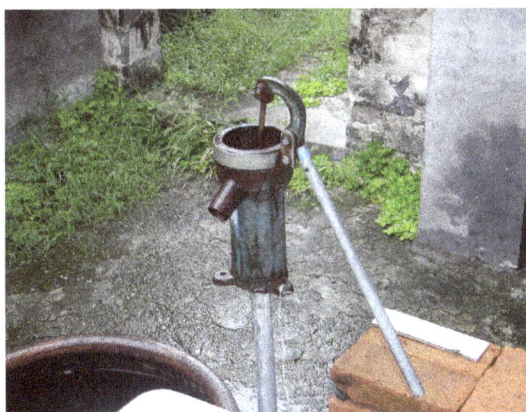

For those who have healthy kidneys, it is rather difficult to drink too much water, but (especially in warm humid weather and while exercising) it is dangerous to drink too little. People can drink far more water than necessary while exercising, however, putting them at risk of water intoxication, which can be fatal. The "fact" that a person should consume eight glasses of water per day cannot be traced back to a scientific source. There are other myths such as the effect of water on weight loss and constipation that have been dispelled.

Original recommendation for water intake in 1945 by the Food and Nutrition Board of the National Research Council read: "An ordinary standard for diverse persons is 1 milliliter for each calorie of food. Most of this quantity is contained in prepared foods." The latest dietary reference intake report by the United States National Research Council in general recommended (including food sources): 2.7 liters of water total for women and 3.7 liters for men. Specifically, pregnant and breastfeeding women need additional fluids to stay hydrated. According to the Institute of Medicine—who recommend that, on average, women consume 2.2 liters and men 3.0 liters—this is recommended to be 2.4 liters (approx. 9 cups) for pregnant women and 3 liters (approx. 12.5 cups) for breastfeeding women, since an especially large amount of fluid is lost during nursing. Also noted is that normally, about 20 percent of water intake comes from food, while the rest comes from drinking water and beverages (caffeinated included). Water is excreted from the body in multiple forms: through urine, feces, sweating, and exhalation of water vapor in the breath. With physical exertion and heat exposure, water loss will increase and daily fluid needs may increase as well.

The single largest freshwater resource suitable for drinking is Lake Baikal in Siberia, which has a very low salt and calcium content and is very clean.

Agriculture

In many developing nations, irrigation accounts for over 90 percent of water withdrawn from available sources for use. In England, where rain is abundant year round, water used for agriculture accounts for less than 1 percent of human usage. Yet even on the same continent, water used for irrigation in Spain, Portugal and Greece exceeds 70 percent of total usage.

Irrigation has been a key component of the "green revolution," which has enabled many developing countries to produce enough food to feed everyone. More water will be needed to produce more food for 3 billion more people. But increasing competition for water and inefficient irrigation practices could constrain future food production.

As a Cleaning Agent

Water is important for washing the human body and everyday items such as clothes, floors, cars, food, and pets.

Standard of Measurement

On April 7, 1795, the gram was defined in France to be equal to "the absolute weight of a volume of pure water equal to a cube of one hundredth of a meter, and to the temperature of the melting ice." For practical purposes though, a metallic reference standard was required, one thousand times more massive, the kilogram. Work was therefore commissioned to determine precisely how massive one liter of water was. In spite of the fact that the decreed definition of the gram specified water at 0 °C—a highly stable temperature point—the scientists chose to redefine the standard and to perform their measurements at the most stable density point: the temperature at which water reaches maximum density, which was measured at the time as 4 °C.

As a Thermal Transfer Agent

Boiling, steaming, and simmering are popular cooking methods that often require immersing food in water or its gaseous state, steam. Water is also used in industrial contexts as a coolant, and in almost all power-stations as a coolant and to drive steam turbines to generate electricity. In the nuclear industry, water can also be used as a neutron moderator.

Recreation

Humans use water for many recreational purposes, as well as for exercising and sports. Some of these include swimming, waterskiing, boating, fishing, and diving. In addition, some sports, like ice hockey and ice skating, are played on ice. Likewise, sports such as skiing or snowboarding require the water to be frozen. Many use water for play fighting, such as with snowballs, water guns, or water balloons.

Lakesides and beaches are popular places for people to go for recreation and relaxation. Many find the sound of flowing water to be calming. Some keep fish and other life in water tanks or ponds for show, fun, and companionship. People also make fountains and use water in their public or private decorations.

Industrial Applications

Pressurized water is used in water blasting and water jet cutters. Also, high-pressure water guns are used for precise cutting. It is also an effective coolant for various machines that generate heat during operation. It works very well, is relatively safe, and is not harmful to the environment.

Food Processing

Water plays many critical roles within the field of food science. Food scientists need to understand the roles of water in food processing, to ensure the success of their products.

Solutes such as salts and sugars found in water affect the physical properties of water. The boiling and freezing points of water is affected by solutes. One mole of sucrose (sugar) raises the boiling point of water by 0.52 °C, and one mole of salt raises the boiling point by 1.04 °C while lowering the freezing point of water in a similar way. Solutes in water also affect water activity which affects many chemical reactions and the growth of microbes in food. Water activity can be described as a ratio of the vapor pressure of water in a solution to the vapor pressure of pure water. Solutes in water lower water activity. This is important to know because most bacterial growth ceases at low levels of water activity. Not only does microbial growth affect the safety of food but also the preservation and shelf life of food.

Water hardness is also a critical factor in food processing. It can dramatically affect the quality of a product as well as playing a role in sanitation. Water hardness is classified based on the amounts of removable calcium carbonate salt it contains per gallon. Water hardness is measured in grains; 0.064 g calcium carbonate is equivalent to one grain of hardness. Water is classified as soft if it contains 1 to 4 grains, medium if it contains 5 to 10 grains and hard if it contains 11 to 20 grains. The hardness of water may be altered or treated by using a chemical ion exchange system. The hardness of water also affects its pH balance which plays a critical role in food processing. For example, hard water prevents successful production of clear beverages. Water hardness also affects sanitation; with increasing hardness, there is a loss of effectiveness for its use as a sanitizer.

Power Generation

Hydroelectricity is electricity obtained from hydropower. Hydroelectric power comes from water driving a turbine connected to a generator. Hydroelectricity is a low-cost, non-polluting, renewable energy source.

Water Resource Distribution and Pollution

Water in itself is not a finite resource (like petroleum is). The water cycle, which involves evaporation, condensation, and precipitation, regenerates potable water in large quantities, many orders of magnitude higher than human consumption. However, many parts of the world are experiencing water scarcity, in the sense that there are problems with the distribution of potable and irrigation water. Such shortages of water form a major social and economic concern and have led to disputes between nations that rely on the same source of water (such as the same river). Some countries experiencing water shortages import water or purify seawater by desalination.

Currently, about 1 billion people around the world routinely drink unhealthy water. Poor water quality and bad sanitation are deadly; some 5 million deaths a year are caused by polluted drinking water.

In the developing world, 90 percent of all wastewater goes untreated into local rivers and streams. Some 50 countries, with roughly a third of the world's population, also suffer from medium or high water stress, and a number of them extract more water annually than is recharged through their natural water cycles. The strain affects surface freshwater bodies like rivers and lakes, but it also degrades groundwater resources.

Water is a strategic resource in the globe and an important element in many political conflicts. Some have predicted that clean water will become the "next oil," making Canada, with this resource in abundance, possibly the richest country in the world. There is a long history of conflict over water, including efforts to gain access to water, the use of water in wars started for other reasons, and tensions over shortages and control.

UNESCO's World Water Development Report from its World Water Assessment Program indicates that, in the next 20 years, the quantity of water available to everyone is predicted to decrease by 30 percent. About 40 percent of the world's inhabitants currently have insufficient fresh water for minimal hygiene. More than 2.2 million people died in 2000 from diseases related to the consumption of contaminated water or drought. In 2004, the UK charity WaterAid reported that a child dies every 15 seconds from easily preventable water-related diseases; often this means lack of sewage disposal; see toilet.

Water Availability in Specific Regions

Ninety-five percent of freshwater in the United States is underground. One crucial source is a huge underground reservoir, the 1,300-kilometer (800 mi) Ogallala aquifer which stretches from Texas to South Dakota and waters one fifth of U.S. irrigated land. Formed over millions of years, the Ogallala aquifer has since been cut off from its original natural sources. It is being depleted at a rate of 12 billion cubic meters (420 billion ft3) per year, amounting to a total depletion to date of a volume equal to the annual flow of 18 Colorado Rivers. Some estimates say it will dry up in as little as 25 years. Many farmers in the Texas High Plains, which rely particularly on the underground source, are now turning away from irrigated agriculture as they become aware of the hazards of over pumping.

The Middle East region has only 1 percent of the worlds available freshwater, which is shared among 5 percent of the world's population. Thus, in this region, water is an important strategic resource. It is predicted that by 2025, countries of the Arabian Peninsula will be using more than twice the amount of water naturally available to them. According to a report by the Arab League, two-thirds of Arab countries have less than 1,000 cubic meters (35,000 ft3) of water per person per year available, which is considered the limit.

Three Gorges Dam, receiving, upstream side

In Asia, Cambodia and Vietnam are concerned about attempts by China and Laos to control the flux of water. China is preparing the Three Gorges Dam project on the Yangtze River, which would become the world's largest dam, causing many social and environmental problems. It also has a project to divert water from the Yangtze to the dwindling Yellow River, which feeds China's most important farming region.

The Ganges is disputed between India and Bangladesh. The water reserves are being quickly depleted and polluted, while the glacier feeding the sacred Hindu river is retreating hundreds of feet each year, causing subsoil streams flowing into the Ganges river to dry up.

In South America, the Guaraní Aquifer is located between the Mercosur countries of Argentina, Brazil, Bolivia and Paraguay. With a volume of about 40,000 km³, it is an important source of fresh potable water for all four countries.

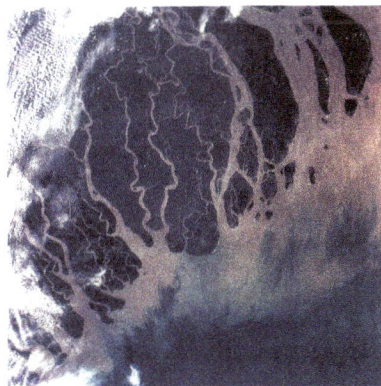

Purification and Waste Reduction

Drinking water is often collected at springs, extracted from artificial borings in the ground, or wells. Building more wells in adequate places is thus a possible way to produce more water, assuming the aquifers can supply an adequate flow. Other water sources are rainwater and river or lake water. This surface water, however, must be purified for human consumption. This may involve removal of undissolved substances, dissolved substances and harmful microbes. Popular methods are filtering with sand which only removes undissolved material, while chlorination and boiling kill harmful microbes. Distillation does all three functions. More advanced techniques are also available, such as reverse osmosis. Desalination of seawater is a more expensive solution, but it is used in some coastal areas with arid climates because the water is abundantly available.

The distribution of drinking water is done through municipal water systems or as bottled water. Governments in many countries have programs to distribute water to the needy at no charge. Others argue that the market mechanism and free enterprise are best to manage this rare resource and to finance the boring of wells or the construction of dams and reservoirs.

Reducing waste by using drinking water only for human consumption is another option. In some cities such as Hong Kong, seawater is extensively used for flushing toilets to conserve freshwater resources.

Polluting water may be the biggest single misuse of water; to the extent that a pollutant limits other uses of the water, it becomes a waste of the resource, regardless of benefits to the polluter. Like other types of pollution, this does not enter standard accounting of market costs, being conceived as externalities for which the market cannot account. Thus other people pay the price of water pollution, while the private firms' profits are not redistributed to the local people who are victims to this pollution. Pharmaceuticals consumed by humans often end up in the waterways and can have detrimental effects on aquatic life if they bioaccumulate.

Fresh Water

Freshwater is water that has little or no dissolved salts and dissolved solids. This excludes sea or marine waters and brackish water. All over the world, water comes in other forms such as ice-sheets, glaciers, lakes, ponds, rivers, streams and icebergs. The quantities found in every geographic area may vary.

Fresh water may be still or fast flowing. Still, fresh water is known as 'Lentic systems' whiles flowing fresh water is known as 'Lotic Systems'. Others come from underground as groundwater in aquifers and underground streams.

Fresh water comes from precipitation from the atmosphere, usually in the form of rain, mist and snow. When these fall, they find their way into streams and rivers which run down from mountain tops to low-lying areas. Eventually, they end up in the sea or ocean. Because much of atmospheric water end up falling into our water bodies, it is important that we keep an eye on the chemicals that find their way into the atmosphere via air pollution.

All freshwater habitats are dominated by the physical properties of water. The molecule is made up of a single oxygen atom with two hydrogen atoms attached. The former is slightly negatively charged whilst the latter is slightly positive. As a consequence, unlike any other solvent on earth, water can attract, dissolve and hold mineral ions in solution. A perfect way of providing nutrients to plants.

Water is viscous and this will produce resistance to animals moving through it whether this is a tiny insect larva, fish or an otter. Hence there may be a need to improve streamlining. Moving water will transport the organism and so those that want to remain in the same area will need to avoid the faster currents. Moving water creates drag on plants trying to anchor themselves. If the current is especially fast silt and mud will be removed and there will be little substrate in which roots can grow. In fact plants may not be evident and a closer inspection (including scraping the rock and placing the debris under the microscope) is needed to see algae and non-flowing plants attached.

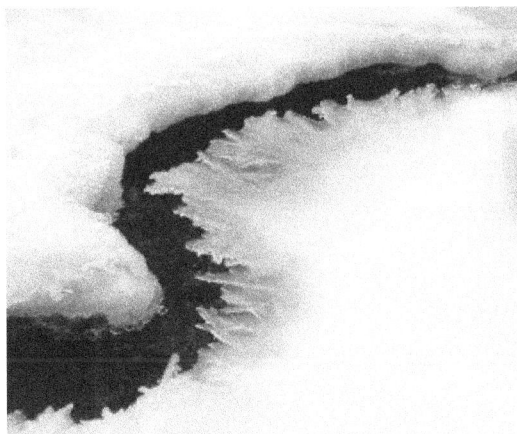

A stream frozen from the edge with the ice on top. The fastest water is still flowing as it has not frozen

The range of temperatures found in the majority of habitats is such that water rarely is found in anything other than liquid. In fact water has a high thermal capacity, meaning that it needs a great deal of energy to heat it but also retains it. This means that living in water creates a stable environment. This stability lasts for much of the year although with the on-set of winter the drop in temperature causes the water to increase its density (as in all liquids). Unusually, water has a useful property in that it is at its most dense at 4 degrees Celcius. A continuing drop in temperature actually decreases the density so that as the water freezes it floats above the denser liquid water. For organisms living in ponds by living at the bottom this may help them survive the winter. Flowing water requires a much lower temperature.

Different Types of Freshwater

Essentially there are two main types, Static Water (called lentic) and Flowing Water (called lotic). However, this is still vague. For example, static water could be any size, a puddle left over from the last rain shower up to Lake Baikal in Russia, the deepest freshwater lake in the world at 1620 metres deep. It also holds the greatest volume of water in the world with over 300 streams feeding it. Incidentally Lake Superior in North America has the greatest surface area.

Lake Baikal

Like lentic systems, flowing water is also varied, from a tiny hill stream bubbling along the edge of a peat bog to a navigable river like the Rhine in Germany. The largest (greatest volume) river system in the world and the second longest is the River Amazon in South America (the Nile is the longest at 6741 kilometres). A few stats: it is navigable for 3700 ks upstream and is 6440 ks at its longest. However, the drainage basin is almost 6 million square kilometres.

Niagra Falls in winter: located between USA and Canada. These are falls on the Canadian side and the mist of water generated freezes on the surrounding vegetation. The water here is cutting back the land at the rate of nearly 1.5 metres a year.

Iguassu Falls: located between Brazil and Argentina. The volume of water flowing over the main cataract (the Devil's Throat, 72 metres high) could fill St Paul's Cathedral in a fraction over 0.5 second. The dark marks in the middle of the spray are swifts that specialise in nesting behind the falls and are flying through it.

Non-natural Water Bodies

These are water created by humans. Typically they will fit within the static water section but some may not be obvious. For example where does a canal fit in? Dug out and created as a transport system several hundred years ago they may now be used for amenities such as narrow boat holidays, canoeing or nature reserves for the wildlife. They need careful management as the process of succession will quickly choke them with vegetation. This is because they fit more to the definition of a pond than a river.

A managed canal

What may appear to be a canal could be a drainage channel in a wetland area or on Dartmoor a leat. These are ways of drawing off water for use in industry or drinking. Either way they are channelled and can be fast flowing like an upland river; quite different to the canal.

Reservoirs are often created by damming the head of a valley and allowing water to collect from the natural process of run-off. They are deep and have more in keeping with a lake. They may be used for amenities, i.e. water sports, but may also be stocked with trout for anglers. Stocking may be the only way of ensuring enough fish as there will usually be insufficient food and unsuitable abiotics for breeding.

Before the latter part of the twentieth century the lakes in Scandinavia were rich in aquatic life including vegetation and a diversity of animals. The invertebrates provided food for trout and salmon but by the 1980's and 1990's this started to become a rarity. In the recent decades sulphur dioxide gases produced by industries burning fossil fuels has been released into the atmosphere. Primarily this has come from countries like the UK. This acidic gas dissolves in the clouds to fall, many miles from where it was produced, as acid rain. This greatly increases the acidity of the lake killing many of the species living there. This is coupled with the release of aluminium from the soil as the acid rain percolates through and drains to the lakes.

Acid Lake in Norway

Sea Water

Seawater is the water that makes up the oceans and seas, covering more than 70 percent of Earth's surface. Seawater is a complex mixture of 96.5 percent water, 2.5 percent salts, and smaller amounts of other substances, including dissolved inorganic and organic materials, particulates, and a few atmospheric gases.

Seawater constitutes a rich source of various commercially important chemical elements. Much of the world's magnesium is recovered from seawater, as are large quantities of bromine. In certain parts of the world, sodium chloride (table salt) is still obtained by evaporating seawater. In addition, water from the sea, when desalted, can furnish a limitless supply of drinking water. Many large desalination plants have been built in dry areas along seacoasts in the Middle East and elsewhere to relieve shortages of fresh water.

Chemical and Physical Properties of Seawater

The six most abundant ions of seawater are chloride (Cl^-), sodium (Na^+), sulfate ($SO_2 4^-$), magnesium ($Mg2^+$), calcium ($Ca2^+$), and potassium (K^+). By weight these ions make up about 99 percent of all sea salts. The amount of these salts in a volume of seawater varies because of the addition or removal of water locally (e.g., through precipitation and evaporation). The salt content in seawater is indicated by salinity (S), which is defined as the amount of salt in grams dissolved in one kilogram of seawater and expressed in parts per thousand. Salinities in the open ocean have been observed to range from about 34 to 37 parts per thousand (0/00 or ppt), which may also be expressed as 34 to 37 practical salinity units (psu).

Inorganic carbon, bromide, boron, strontium, and fluoride constitute the other major dissolved substances of seawater. Of the many minor dissolved chemical constituents, inorganic phosphorus and inorganic nitrogen are among the most notable, since they are important for the growth of organisms that inhabit the oceans and seas. Seawater also contains various dissolved atmospheric gases, chiefly nitrogen, oxygen, argon, and carbon dioxide. Some other components of seawater are dissolved organic substances, such as carbohydrates and amino acids, and organic-rich particulates. These materials originate primarily in the upper 100 metres (330 feet) of the ocean, where dissolved inorganic carbon is transformed by photosynthesis into organic matter.

Many of the characteristics of seawater correspond to those of water in general, because of their common chemical and physical properties. For example, the molecular structure of seawater, like that of fresh water, favours the formation of bonds among molecules. Some of the distinctive qualities of seawater are attributable to its salt content. The viscosity (i.e., internal resistance to flow) of seawater, for example, is higher than that of fresh water because of its higher salinity. The density of seawater also is higher for the same reason. Seawater's freezing point is lower than that of pure water, and its boiling point is higher.

Chemical Composition

The chemical composition of seawater is influenced by a wide variety of chemical transport mechanisms. Rivers add dissolved and particulate chemicals to the oceanic margins. Wind-borne particulates are carried to mid-ocean regions thousands of kilometres from their continental source areas. Hydrothermal solutions that have circulated through crustal materials beneath the seafloor add both dissolved and particulate materials to the deep ocean. Organisms in the upper ocean convert dissolved materials to solids, which eventually settle to greater oceanic depths. Particulates in transit to the seafloor, as well as materials both on and within the seafloor, undergo chemical exchange with surrounding solutions. Through these local and regional chemical input and removal mechanisms, each element in the oceans tends to exhibit spatial and temporal concentration variations. Physical mixing in the oceans (thermohaline and wind-driven circulation) tends to homogenize the chemical composition of seawater. The opposing influences of physical mixing and of biogeochemical input and removal mechanisms result in a substantial variety of chemical distributions in the oceans.

Dissolved Inorganic Substances

The principal components of seawater are listed in the table.

Principal constituents of seawater*			
ionic constituent	g/kg of seawater	moles/kg**	moles/kg**
chloride	19.162	0.5405	1.0000
sodium	10.679	0.4645	0.8593
magnesium	1.278	0.0526	0.0974
sulfate	2.680	0.0279	0.0517
calcium	0.4096	0.01022	0.0189
potassium	0.3953	0.01011	0.0187
carbon (inorganic)	0.0276	0.0023	0.0043

bromide	0.0663	0.00083	0.00154
boron	0.0044	0.00041	0.00075
strontium	0.0079	0.00009	0.000165
fluoride	0.0013	0.00007	0.000125

In contrast to the behavior of most oceanic substances, the concentrations of the principal inorganic constituents of the oceans are remarkably constant. Calculations indicate that, for the main constituents of seawater, the time required for thorough oceanic mixing is quite short compared with the time that would be required for input or removal processes to significantly change a constituent's concentration. The concentrations of the principal constituents of the oceans vary primarily in response to a comparatively rapid exchange of water (precipitation and evaporation), with relative concentrations remaining nearly constant.

Salinity is used by oceanographers as a measure of the total salt content of seawater. Practical salinity, symbol S, is determined through measurements of the electrical conductivity and temperature of seawater, which are interpreted by an algorithm developed by the United Nations Educational Scientific and Cultural Organization (UNESCO). Practical salinity, along with temperature, can be used to calculate precisely the density of seawater samples. Because of the constant relative proportions of the principal constituents, salinity can also be used to directly calculate the concentrations of the major ions in seawater. The measure of practical salinity was originally developed to provide an approximate measure of the total mass of salt in one kilogram of seawater. Seawater with S equal to 35 contains approximately 35 grams of salt and 965 grams of water, or 35 ppt (35 psu).

Many other constituents are of great importance to the biogeochemistry of the oceans. Such chemicals as inorganic phosphorus (HPO_4^{2-} and PO_4^{3-}) and inorganic nitrogen (NO_3^-, NO_2^-, and NH_4^+) are essential to the growth of marine organisms. Nitrogen and phosphorus are incorporated into the tissues of marine organisms in approximately a 16:1 ratio and are eventually returned to solution in approximately the same proportion. As a consequence, in much of the oceanic waters dissolved inorganic phosphorus and nitrogen exhibit a close covariance. Dissolved inorganic phosphorus distributions in the Pacific Ocean strongly bear the imprint of phosphorus incorporation by organisms in the surface waters of the ocean and of the return of the phosphorus to solution via a rain of biological debris remineralized in the deep ocean. Inorganic phosphate concentrations in the western Pacific range from somewhat less than 0.1 micromole/kg ($1 \times 10-7$ mole/kg) at the surface to approximately 3 micromoles/kg ($3 \times 10-6$ mole/kg) at depth. Inorganic nitrogen ranges between somewhat less than 1 micromole/kg and 45 micromoles/kg along the same section of ocean and exhibits a striking covariance with phosphate.

A variety of elements essential to the growth of marine organisms, as well as some elements that have no known biological function, exhibit nutrient-like behavior broadly similar to nitrate and phosphate. Silicate is incorporated into the hard structural parts of certain types of marine organisms (diatoms and radiolarians) that are abundant in the upper ocean.

Dissolved Organic Substances

Processes involving dissolved and particulate organic carbon are of central importance in shaping the chemical character of seawater. Marine organic carbon principally originates in the uppermost

100 metres of the oceans where dissolved inorganic carbon is photo synthetically converted to organic materials. The "rain" of organic-rich particulate materials, resulting directly and indirectly from photosynthetic production, is a principal factor behind the distributions of many organic and inorganic substances in the oceans. A large fraction of the vertical flux of materials in the uppermost waters is converted to dissolve substances within the upper 400 metres (about 1,300 feet) of the oceans. Dissolved organic carbon (DOC) accounts for at least 90 percent of the total organic carbon in the oceans. Estimates of DOC appropriate to the surface of the open ocean range between roughly 100 and 500 micromoles of carbon per kilogram of seawater. DOC concentrations in the deep ocean are 5 to 10 times lower than surface values. DOC occurs in an extraordinary variety of forms, and, in general, its composition is controversial and poorly understood. Conventional techniques have indicated that, in surface waters, about 15 percent of DOC can be identified as carbohydrates and combined amino acids. At least 1–2 percent of DOC in surface waters occurs as lipids and 20–25 percent as relatively unreactive humic substances. The relative abundances of reactive organic substances, such as amino acids and carbohydrates, are considerably reduced in deep ocean waters. Dissolved and particulate organic carbon in the surface ocean participates in diel cycles (i.e., those of a 24-hour period) related to photosynthetic production and photochemical transformations. The influence of dissolved organic matter on ocean chemistry is often out of proportion to its oceanic abundance. Photochemical reactions involving DOC can influence the chemistry of vital trace nutrients such as iron, and, even at dissolved concentrations on the order of one nanomole/kg (1×10–9 mole/kg), dissolved organic substances in the upper ocean waters are capable of greatly altering the bioavailability of essential trace nutrients, as, for example, copper and zinc.

Effects of Human Activities

Although the oceans constitute an enormous reservoir, human activities have begun to influence their composition on both a local and a global scale. The addition of nutrients (through the discharge of untreated sewage or the seepage of soluble mineral fertilizers, for example) to coastal waters results in increased phytoplankton growth, high levels of dissolved and particulate organic materials, decreased penetration of light through seawater, and alteration of the community structure of bottom-dwelling organisms. Through industrial and automotive emissions, lead concentrations in the surface ocean have increased dramatically on a global scale compared with pre-industrial levels. Certain toxic organic compounds, such as polychlorinated biphenyls (PCBs), are found in seawater and marine organisms and are attributable solely to the activities of humankind. Although most radioactivity in seawater is natural (approximately 90 percent as potassium-40 and less than 1 percent each as rubidium-87 and uranium-238), strontium-90 and certain other artificial radioisotopes have unique environmental pathways and potential for bioaccumulation (that is, concentration in higher levels of the food chain).

Among the most dramatic influences of human activities on a global scale is the remarkable increase of carbon dioxide levels in the atmosphere. Atmospheric carbon dioxide levels are expected to surpass 420 parts per million by volume by the middle of the 21st century, with potentially profound consequences for global climate and agricultural patterns. It is thought that the oceans, as a great reservoir of carbon dioxide, will ameliorate this consequence of human activities to some degree. However, ocean acidification due to the absorption of carbon dioxide is an emerging environmental problem.

Salinity Distribution

A discussion of salinity, the salt content of the oceans, requires an understanding of two important concepts: (1) the present-day oceans are considered to be in a steady state, receiving as much salt as they lose, and (2) the oceans have been mixed over such a long time period that the composition of sea salt is the same everywhere in the open ocean. This uniformity of salt content results in oceans in which the salinity varies little over space or time.

The range of salinity observed in the open ocean is from 33 to 37 grams of salt per kilogram of seawater or psu. For the most part, the observed departure from a mean value of approximately 35 psu is caused by processes at Earth's surface that locally add or remove fresh water. Regions of high evaporation have elevated surface salinities, while regions of high precipitation have depressed surface salinities. In nearshore regions close to large freshwater sources, the salinity may be lowered by dilution. This is especially true in areas where the region of the ocean receiving the fresh water is isolated from the open ocean by the geography of the land.

Areas of the Baltic Sea may have salinity values depressed to 10 psu or less. Increased salinity by evaporation is accentuated where isolation of the water occurs. This effect is found in the Red Sea, where the surface salinity rises to 41 psu. Coastal lagoon salinities in areas of high evaporation may be much higher. The removal of fresh water by evaporation or the addition of fresh water by precipitation does not affect the constancy of composition of the sea salt in the open sea. A river draining a particular soil type, however, may bring to the oceans only certain salts that will locally alter the salt composition. In areas of high evaporation where the salinity is driven to very high values, precipitation of particular salts may alter the composition too. At high latitudes where sea ice forms seasonally and icebergs are often released into the open ocean, the salinity of the seawater is reduced when ice melts and is elevated during ice formation. This saltier water can then sink down into the deep ocean.

At depth in the oceans, salinity may be altered as seawater percolates into fissures associated with deep-ocean ridges and crustal rifts involving volcanism. This water then returns to the ocean as superheated water carrying dissolved salts from the magmatic material within the crust. It may lose much of its dissolved load to precipitates on the seafloor and gradually blend in with the surrounding seawater, sharing its remaining dissolved substances.

Salt concentrations as high as 256 psu have been found in hot but dense pools of brine trapped in depressions at the bottom of the Red Sea. The composition of the salts in these pools is not the same as the sea salt of the open oceans.

The salinities found at the greater depths of the open oceans are quite uniform in both time and space with average values of 34.5 to 35 psu. These salinities are determined by surface processes such as those described above when the water, now at depth, was last in contact with the surface.

The intertropical convergence, with its high precipitation centred about 5° N, supports the tropical rainforests of the world and leaves its imprint on the oceans as a latitudinal depression of surface salinity. At approximately 30°–35° N and 30°–35° S, the subtropical zones called the horse latitudes are belts of high evaporation that produce major deserts and grasslands on the continents and cause the surface salinity to rise. At 50°–60° N and 50°–60° S, precipitation again increases.

Ground Water

Groundwater is the water beneath the surface of the ground in the zone of saturation where every pore space between rock and soil particles is saturated with water. Above the zone of saturation is an area where both air and moisture are found in the spaces between soil and rock particles. This is called the zone of aeration. Water percolates (moves downward) through this zone until it reaches the zone of saturation. The water table is the top of the saturated zone.

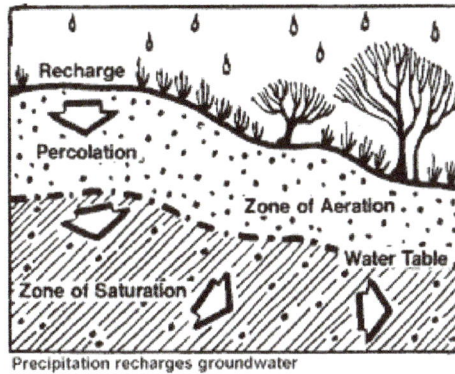

Precipitation recharges groundwater

Origin of Groundwater

It's not as mysterious as it seems. The real sources of groundwater are rain and snow. Rain and melting snow percolate into the ground and saturate the pores between rock and soil particles. Geologists call this process groundwater recharge and, the places where it occurs, recharge areas.

Where Does it Go

Once it reaches the zone of saturation under the ground, groundwater begins to move slowly by the force of gravity through the interconnecting pore spaces until it reaches a discharge area, where it seeps or flows out into a wetland, spring, river, or pond to become part of the surface water.

Water evaporates from surface water bodies and from land surfaces and returns to the atmosphere. Plants transpire water into the atmosphere. Water in the atmosphere condenses into rain. Some of the rain recharges the groundwater, and the cycle keeps repeating. Groundwater, in other words, is part of the hydrologic cycle. Groundwater and surface water are interconnected; groundwater becomes surface water when it discharges to surface water bodies. Most streams keep flowing during the dry summer months because groundwater discharges into them from the zone of saturation - this flow is called baseflow. Under certain conditions the flow may be reversed and the surface water may recharge the groundwater. Only a portion of the water that falls as rain or snow in Massachusetts actually recharges the groundwater. The rest runs off into surface water bodies, is taken up by plants and transpired, or evaporates.

Groundwater Movement

Groundwater is always moving from higher recharge areas to lower discharge areas; however, it moves slowly. Groundwater movement is measured in feet per day or, in some cases, in feet per year. In contrast, surface water movement is measured in feet per second. The speed at which groundwater moves are determined by the types of material it must flow through and the steepness of the gradient from the recharge area to discharge area. Water moves more easily through the large pores of sand and gravel, for example, than through material that contains fine silt and clay.

The water table is at the top of the zone of saturation, but it doesn't remain at one level all the time. The rise and fall of the water table is a natural part of the groundwater system. It occurs seasonally each year. In the late winter and early spring (February, March, April), melting snow and rain percolate into the ground to raise the water table to its annual high level. During the growing season, rainwater is used by plants for transpiration or it evaporates. As a result, little or no groundwater recharge occurs during the late spring and summer months. During that time, however, groundwater continues to discharge into streams, lakes, and wetlands, so the water table drops. By fall (October and November), the water table has dropped as much as fifteen feet to its lowest annual level. The groundwater is recharged again by rain that falls after the growing season. There is no recharge in the winter when the ground is frozen, but recharge can occur during midwinter thaws. During the winter, water is stored in the snow pack. In the spring, the melting snow recharges the groundwater, raising the water table to its annual high level again.

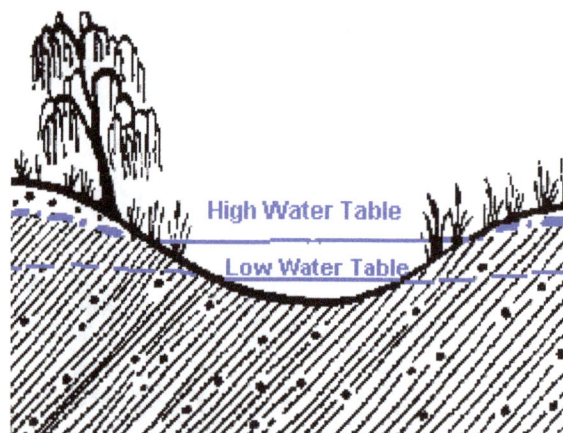

Water Cycle

The water cycle describes how water is exchanged (cycled) through Earth's land, ocean, and atmosphere. Water always exists in all three places, and in many forms—as lakes and rivers, glaciers and ice sheets, oceans and seas, underground aquifers, and vapor in the air and clouds.

Evaporation, Condensation, and Precipitation

The water cycle consists of three major processes: evaporation, condensation, and precipitation.

Evaporation

Evaporation is the process of a liquid's surface changing to a gas. In the water cycle, liquid water (in the ocean, lakes, or rivers) evaporates and becomes water vapor.

Water vapor surrounds us, as an important part of the air we breathe. Water vapor is also an important greenhouse gas. Greenhouse gases such as water vapor and carbon dioxide insulate the Earth and keep the planet warm enough to maintain life as we know it.

The water cycle's evaporation process is driven by the sun. As the sun interacts with liquid water on the surface of the ocean, the water becomes an invisible gas (water vapor). Evaporation is also influenced by wind, temperature, and the density of the body of water.

Condensation

Condensation is the process of a gas changing to a liquid. In the water cycle, water vapor in the atmosphere condenses and becomes liquid.

Condensation can happen high in the atmosphere or at ground level. Clouds form as water vapor condenses, or becomes more concentrated (dense). Water vapor condenses around tiny particles called cloud condensation nuclei (CCN). CCN can be specks of dust, salt, or pollutants. Clouds at ground level are called fog or mist.

Like evaporation, condensation is also influenced by the sun. As water vapor cools, it reaches its saturation limit, or dew point. Air pressure is also an important influence on the dew point of an area.

Precipitation

Unlike evaporation and condensation, precipitation is not a process. Precipitation describes any liquid or solid water that falls to Earth as a result of condensation in the atmosphere. Precipitation includes rain, snow, and hail.

Fog is not precipitation. The water in fog does not actually precipitate, or liquify and fall to Earth. Fog and mist are a part of the water cycle called suspensions: They are liquid water suspended in the atmosphere.

Precipitation is one of many ways water is cycled from the atmosphere to the Earth or ocean.

Other Processes

Evaporation, condensation, and precipitation are important parts of the water cycle. However, they are not the only ones.

Runoff, for instance, describes a variety of ways liquid water moves across land. Snowmelt, for example, is an important type of runoff produced as snow or glaciers melt and form streams or pools.

Transpiration is another important part of the water cycle. Transpiration is the process of water vapor being released from plants and soil. Plants release water vapor through microscopic pores called stomata. The opening of stomata is strongly influenced by light, and so is often associated with the sun and the process of evaporation. Evapotranspiration is the combined components of evaporation and transpiration, and is sometimes used to evaluate the movement of water in the atmosphere.

States of Water

Through the water cycle, water continually circulates through three states: solid, liquid, and vapor.

Ice is solid water. Most of Earth's freshwater is ice, locked in massive glaciers, ice sheets, and ice caps.

As ice melts, it turns to liquid. The ocean, lakes, rivers, and underground aquifers all hold liquid water.

Water vapor is an invisible gas. Water vapor is not evenly distributed across the atmosphere. Above the ocean, water vapor is much more abundant, making up as much as 4% of the air. Above isolated deserts, it can be less than 1%.

The Water Cycle and Climate

The water cycle has a dramatic influence on Earth's climate and ecosystems.

Climate is all the weather conditions of an area, evaluated over a period of time. Two weather conditions that contribute to climate include humidity and temperature. These weather conditions are influenced by the water cycle.

Humidity is simply the amount of water vapor in the air. As water vapor is not evenly distributed by the water cycle, some regions experience higher humidity than others. This contributes to radically different climates. Islands or coastal regions, where water vapor makes up more of the atmosphere, are usually much more humid than inland regions, where water vapor is scarcer.

A region's temperature also relies on the water cycle. Through the water cycle, heat is exchanged and temperatures fluctuate. As water evaporates, for example, it absorbs energy and cools the local environment. As water condenses, it releases energy and warms the local environment.

The Water Cycle and the Landscape

The water cycle also influences the physical geography of the Earth. Glacial melt and erosion caused by water are two of the ways the water cycle helps create Earth's physical features.

As glaciers slowly expand across a landscape, they can carve away entire valleys, create mountain peaks, and leave behind rubble as big as boulders. Yosemite Valley, part of Yosemite National Park in the U.S. state of California, is a glacial valley. The famous Matterhorn, a peak on the Alps between Switzerland and Italy, was carved as glaciers collided and squeezed up the earth between them. Canada's "Big Rock" is one of the world's largest "glacial erratics," boulders left behind as a glacier advances or retreats.

Glacial melt can also create landforms. The Great Lakes, for example, are part of the landscape of the Midwest of the United States and Canada. The Great Lakes were created as an enormous ice sheet melted and retreated, leaving liquid pools.

The process of erosion and the movement of runoff also create varied landscapes across the Earth's surface. Erosion is the process by which earth is worn away by liquid water, wind, or ice.

Erosion can include the movement of runoff. The flow of water can help carve enormous canyons, for example. These canyons can be carved by rivers on high plateaus (such as the Grand Canyon, on the Colorado Plateau in the U.S. state of Arizona). They can also be carved by currents deep in the ocean (such as the Monterey Canyon, in the Pacific Ocean off the coast of the U.S. state of California).

Reservoirs and Residence Time

Reservoirs are simply where water exists at any point in the water cycle. An underground aquifer can store liquid water, for example. The ocean is a reservoir. Ice sheets are reservoirs. The atmosphere itself is a reservoir of water vapor.

Residence time is the amount of time a water molecule spends in one reservoir. For instance, the residence time of "fossil water," ancient groundwater reservoirs, can be thousands of years. Some fossil water reservoirs beneath the Sahara Desert have existed for 75,000 years.

Residence time for water in the Antarctic ice sheet is about 20,000 years. That means that a molecule of water will stay as ice for about that amount of time.

The residence time for water in the ocean is much shorter—about 3,200 years.

The residence time of water in the atmosphere is the shortest of all—about nine days.

Calculating residence time can be an important tool for developers and engineers. Engineers may consult a reservoir's residence time when evaluating how quickly a pollutant will spread through the reservoir, for instance. Residence time may also influence how communities use an aquifer.

Water Chemistry

The chemistry of water deals with the fundamental chemical property and information about water. Water chemistry is discussed in the following subtitles.

- Composition of water.
- Structure and bonding of water.
- Molecular Vibration of water.
- Symmetry of water molecules.
- Formation of hydrogen bonding in water.

- Structure of ice.

- Autoionization.

- Leveling effect of water and acid-base characters.

- Amphiprotic nature.

- Reactivity of water towards alkali metals; alkaline earth metals; halogens; hydrides; methane; oxides; and oxygen ions.

- Electrolysis of water.

Composition of Water

Water consists of only hydrogen and oxygen. Both elements have natural stable and radioactive isotopes. Due to these isotopes, water molecules of masses roughly $18\,(H_2{}^{16}O)$ to $22\,(D_2{}^{18}O)$ are expected to form. Isotopes and their abundances of H and O are given below. From these data, we can estimate the relative abundances of all isotopic water molecules.

Abundances (% or halflife) of hydrogen and oxygen isotopes

H	^{2}D	^{3}T		
99.985%	0.015%	12.33 y		
^{14}O	^{15}O	^{16}O	^{17}O	^{18}O
70.6 s	122 s	99.762%	0.038%	0.200%

Relative abundance of isotopic water

$H_2{}^{16}O$	$H_2{}^{18}O$	$H_2{}^{17}O$	$HD^{16}O$	$D_2{}^{16}O$	$HT^{16}O$
99.78%	0.20%	0.03%	0.0149%	0.022 ppm	trace
18	20	19	19	20	20 amu

The predominant water molecules $H_2{}^{16}O$ have a mass of 18 amu, but molecules with mass 19 and 20 occur significantly. Because the isotopic abundances are not always the same due to their astronomical origin, the isotopic distribution of water molecules depends on its source and age. Its study is linked to other sciences. Chemistry of Water and Water Pollution, Ellis Harwood for isotopic distribution of water.

In particular, $D_2{}^{16}O$ is called heavy water, and it is produced by enrichment from natural water. Properties of heavy water are particularly interesting due to its application in nuclear technology.

Structure and Bonding of the Water Molecule

Pure water, H_2O, has a unique molecular structure. The O-H bond lengths are 0.096 nm and the H-O-H angle = 104.5°. This strange geometry can be explained by various methods.

```
                Lewis Dot Structures

        H                H
        |                |            ..               ..
   H--C--H          H--N :      H--O :          H--F :
        |                |            |
        H                H            H                ..

      CH₃              NH₃          H₂O             HF

                   Bondlength /pm

      C-H              N-H          O-H             H-F
      109              101          96              92
```

From carbon to neon, the numbers of valence electrons increase from 4 to 8. These elements require 4, 3, 2, 1, and 0 H atoms to share electrons in order to complete the octet requirement. Their Lewis dot structures are shown on the right, and note the trend in bond lengths.

There are six valance electrons on the oxygen, and one each from the hydrogen atom in the water molecule. The eight electrons form two H-O bonds, and left two lone pairs. The long pairs and bonds stay away from each other and they extend towards the corners of a tetrahedron. Such an ideal structure should give H-O-H bond angle of 109.5°, but the lone pairs repel each other more than they repel the O-H bonds. Thus, the O-H bonds are pushed closer, making the H-O-H angle less than 109°.

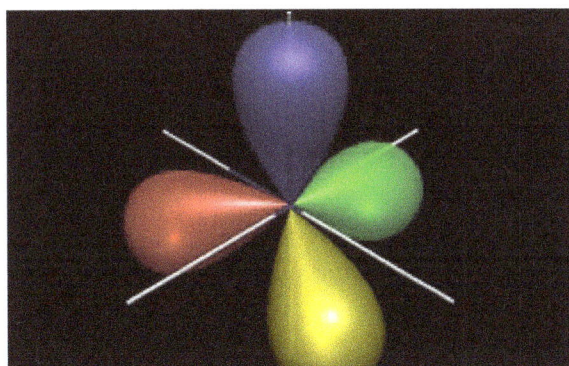

After the introduction of quantum mechanics, the electronic configuration for the valence electron of oxygen are $2s^2 2p^4$. Since the energy levels of 2_s and 2_p are close, valence electrons have characters of both s and p. The mixture is called sp^3 hybridization. These hybridized orbitals are shown on the right. The structures of CH_4, NH_3, and H_2O can all explained by these hybrid orbitals of the central atoms. The above approach is the valence bond theory, and both the C-H bonds and lone electron pairs are counted as VSPER pairs in the Valence-shell Electron-Pair Repulsion (VSEPR) model, according to which, the four groups point to the corners of a tetrahedron.

For triatomic molecules such as water, molecular orbital (MO) approach can also be applied to discuss the bonding. The result however is similar to the valence bond approach, but the MO theory gives the energy levels of the electron for further exploration.

Molecular Vibration of Water

Atoms in a molecule are never at rest, and for each type of molecule, there are some normal vibration modes. For the water molecule, the three normal modes of vibrations are symmetric stretching, bending and asymmetric stretching.

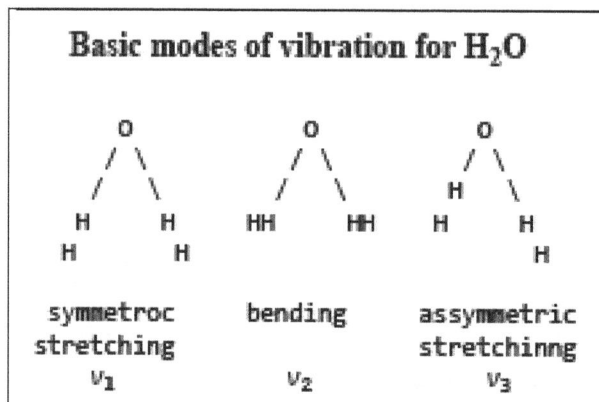

Basic modes of vibration for H_2O

symmetroc stretching v_1 — bending v_2 — assymmetric stretchinng v_3

The vibrations are quantized, as do any microscopic system, and their quantum numbers are designated as v_1, v_2 and v_3. The observed transition bands of D_2O, H_2O, and HDO are given in the table.

Transition bands of D_2O, H_2O, and HDO					
Quantum numbers of upper state			Absorption wavenumbers of bands /cm-1		
v_1	v_2	v_3	D_2O	H_2O	HDO
0	1	0	1178	1594	1402
1	0	0	2671	3656	2726
0	0	1	2788	3756	3703
0	1	1	3956	5332	5089

Data from Eisenberg, D. and Kauzmann, W. (1969) Structure and properties of water, Oxford University press.

The ideal transition bands are centered in the given wavenumbers. However, these wavenumbers are calculated based on isolated molecules with no interaction with any neighbours. When molecules interact with each other, the energy levels are modified, and the bands shift.

Many more less intense absorption bands extend into the green part of the visible spectrum. The absorption spectrum of water may contribute to the blue color for lake, river and ocean waters.

Symmetry of Water Molecules

The water molecules are rather symmetric in that there are two mirror planes of symmetry, one containing all three atoms and one perpendicular to the plane passing through the bisector of the H-O-H angle. Furthermore, if the molecules are rotated 180° (360°/2) the shape of the molecule is unperturbed. This indicates that the molecules have a 2-fold rotation axis. The three symmetry elements are 2-fold rotation, and two mirror planes. Both mirror planes contain the rotation axis, and this type of symmetry belongs to the point group C2v.

A point group has a definite number of symmetry elements arranged in certain fashion. Molecules can be classified according to their point groups. Molecules of the same point group have similar spectroscopic characters. Other molecules of C2v point group are, the $CH_2 = O$, CH_2Cl_2, the bent O_3 etc.

Formation of Hydrogen Bonding

Under certain conditions, an atom of hydrogen is attracted by rather strong forces to two atoms instead of only one, so that it may be considered to be acting as a bond between them. This is called hydrogen bond. This statement is from Linus Pauling in his book The Nature of the Chemical Bond. He gave the ion [F:H:F]- as an example. At that time, the hydrogen bond was recognized as mainly ionic in nature. The energy associated with hydrogen bond is 8 to 40 kJ/mol.

Normally, the melting point and boiling point of a substance increase with molecular mass. For example the melting points of inert gases are 0.95, 24.48, 83.8, and 116.6 K respectively for He, Ne, Ar, and Kr.

In this table, the melting and boiling points for water are particular high for its small molecular mass. This is usually attributed to the formation of hydrogen bonds. The small electronegative atoms F, O and N are somewhat negatively charged when they are bonded to hydrogen atoms. The negative charges on F, O and N attract the slightly positive hydrogen atoms, forming a strong interaction called hydrogen bond.

Comparison of melting and boiling points for a few substances			
Molecule	Molar mass	m.p.	b.p. /° C
NH_3	17	-77.8	-33.5
H_2O	18	-0	100
H_2S	34	-85.6	-60
H_2Se	81	-60.4	-41.5
H_2Te	128.6	-51	-1.8
C_2H_5OH	32	?	65
C_2H_5OH	46	?	78
$C_2H_5OC_2H_5$	74	?	34

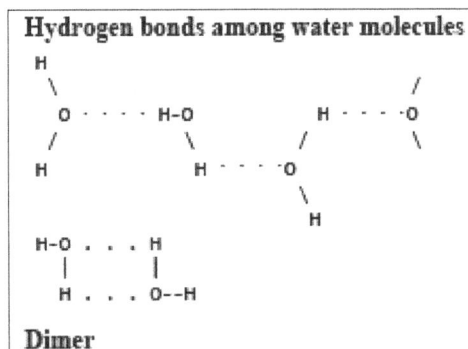

Hydrogen bonds among water molecules

Dimer

A graph showing the melting points and boiling points of group 16 provided by Prof. J. Boucher illustrates the same point.

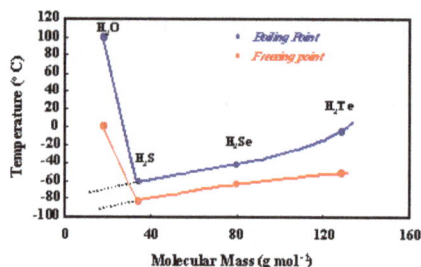

**Boiling and Freezing Points
of Group 16 Hydrides**

Based on the observed absorption at 3546 and 3546 and 3691 cm^{-1}, Van Thiel, Becker, and Pinmentel suggested the formation of water dimer when trapped in a matrix of nitrogen.

Due to hydrogen bonding, water molecules form dimers, trimers, polymers, and clusters. The hydrogen bonds are not necessarily liner.

Structure of Ice

Ice occurs in many places, including the Antarctic. If all the ice melted, the water level of the oceans will rise about 70 m. The structure of ice and the caption are from this link.

The density of ice is dramatically smaller than that of water, due to the regular arrangement of water molecule via hydrogen bonds. In an idealized structure of ice, every hydrogen atom is involved in hydrogen bond. Every oxygen atom is surrounded by four hydrogen bonds.

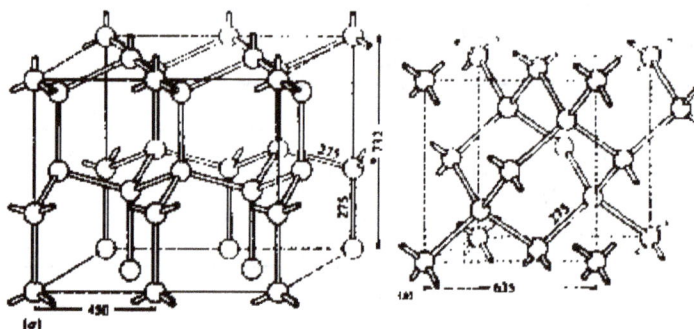

The diagram shows the structure of hexagonal ice in (a) and cubic ice in (b). A rod here represents a hydrogen bond. Since the hydrogen bonds are not linear, the real structure is a little more complicated.

The tetrahedral coordination opens up the space between molecules. On each hydrogen bond, shown by a rod joining the oxygen atoms, lies one proton in an asymmetric position. Bond lengths, 275 pm, are indicated. Ordinary ice is hexagonal. And the hexagonal c axis is labelled 732 pm, and one of the hexagonal an axes is labelled 450 pm. If water vapor condenses on very cold substrate at

143-193 K (-130 to -80°C) a cubic phase is formed. In (b) the cubic unit cell is outlined with dashed lines; dimensions are in pm determined at 110 K.

These diagrams can also be used to represent the two forms of diamond, and in this case, the rods joining the atoms represent C-C bonds. Each C-C bondlength is 154 pm. Silicon and germanium crystals have the same structure, but their bondlengths are longer. The two diamond types of structure are related to the packing of spheres. The hexagonal type has the ABABAB... sequence, whereas the cubic type has the ABCABC... sequence. In both cases, half of the tetrahedral sites are occupied by tetrahedrally bonded carbon atoms. Hexagonal diamonds have been observed in meteorites.

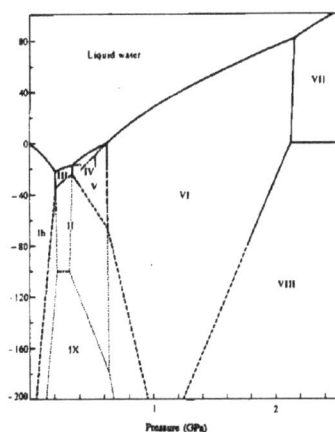

The four hydrogen bonds around an oxygen atom form a tetrahedron in a fashion found in the two types of diamonds. Thus, ice, diamond, and close packing of spheres are somewhat topologically related.

A phase diagram of water shows 9 different solid phases (ices). Ice Ih is the ordinary ice. In addition to ice Ic from vapor deposition, conditions for nine phases are shown. Aside from ice I, other phases are formed and observed under high pressure generated by machines built by scientists. So far, ten different forms of ice have been observed, and some ice forms exist at very high pressure. The pressure deep under the polar (Antarctic) ice cap is very high, but we are not able to make any direct observation or study.

There is a report of the 11th ice, and the ice phase diagram and drawings of ice structures given here is extremely interesting.

The Autoionization of Water

The Autoionization of Water in the formation of ions according to:

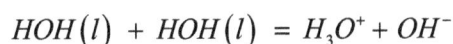

$$HOH(l) + HOH(l) = H_3O^+ + OH^-$$

This is an equilibrium process and is characterised by an equilibrium constant, K'w:

$$K'_w = \frac{[H_3O^+][OH^-]}{[H_2O]}$$

Since $[H_2O]$ = 1000/18 = 55.56 M, and remains rather constant under any circumstance, we usually write,

$$K_w = [H_3O^+][OH^-]$$
$$= 10^{-14} \text{ (or 1e-14)}$$
$$pK_w = -\log K_w \quad \text{(defined)}$$
$$= 14 \text{ (at 298 K)}$$

For neutral water, [H3O+] = [OH-] = 1e-7 at this temperature. Furthermore, we define,

$$pH = -\log[H_3O^+]$$

$$pOH = -\log[OH^-]$$

$$pH = pOH = 7 \text{ at } 298 \text{ K}; \quad (\text{in neutral solutions})$$

It is important to realize that Kw depends on temperature.

Leveling Effect of Water and Acid-base Characters

The strength of strong acids and bases is dominated by the autoionization of water. In aqueous solutions, the strongest acid and base are the hydronium ion, H_3O^+, and the hydroxide ion OH^- respectively. Acids HCl, HBr, HI, HNO_3, $HClO_3$, $HClO_4$, and H_2SO_4 completely ionize in water, making them as strong as H_3O^+ due to the leveling effect of water. Furthermore, strong acids, strong bases, and salts completely ionize in their aqueous solutions.

For example, HCl is a stronger acid than H_2O, and the reaction takes place as HCl dissolves in water.

$$HCl + H_2O = Cl^- + H_3O^+$$

A similar equation can be written for another strong acid.

On the other hand, a strong base also react with water to give the stong base species, OH^-.

$$H_2O + B^- = OH^- + HB$$

For example, O^{2-}, CH_3O^-, and NH_2 are strong bases. The leveling effect also apply to bases.

Amphiprotic Species

Equilibria of acids and bases are interesting chemistry. When an acid and a base differ by a proton, they are called a conjugate acid-base pair. A water molecule is a weak acid and base, due to its ability to accept or donate a proton. Such properties make water an amphiprotic species. In fact, H_3O^+, H_2O and OH^- are amphiprotic, as are some other conjugate acid-base pairs of weak acids and bases.

If several acids and bases are dissolved in water, all equilibria must be considered. To estimate the pH of these solutions requires the exact treatment of several equilibrium constants. For example, many species dissolve in rain water, and many equilibria must be considered. Detail consideration and examples are given in Acid-Base Reactions.

Carbon dioxide in the air dissolve in rain water, lakes and rivers. A solution of CO_2 involves the following reaction:

Reaction	K formula	K value
$H_2O(l) + CO_2(g) = H_2CO_3(l)$	$1/P_{CO2}$	?
$H_2CO_3 = HCO_3^- + H^+$	$[HCO_3^-] [H^+] / [H_2CO_3]$	5e-7
$HCO_3^- = CO_3^{-2} + H^+$	$[CO_3^{-2}] [H^+] / [HCO_3^-]$	5e-11
$HOH(l) + HOH(l) = H_3O^+ + OH^-$	$[H_3O^+] [OH^-]$	1e-14

These complicated equilibria make natural water a buffer.

Reactivity of Water Towards Metals

Alkali metals react with water readily. Contact of cesium metal with water causes immediate explosion, and the reactions become slower for potassium, sodium and lithium. Reactions with barium, strontium, calcium are less well known, but they do react readily. Warm water may be needed to react with calcium metal, however.

Many metals displace H^+ ions in acidic solutions. This is often seen as a property of acids.

Electrolysis of Water

The enthalpy of formation for liquid water, $H_2O(l)$, is -285.830 and that of water vapor is -241.826 kJ / mol. The difference is the heat of vaporization at 298 K. Liquid water and vapor entropies (S) are 69.95 and 188.835 kJ K^{-1} mol^{-1} respectively. These are entropies, not standard entropies of formation. The entropy of formation for water is obtained by,

$$\Delta DS^o_{f\ water} = S^o_{water} - S^o_{H_2} - 0.5 S^o_{O_2}$$

$$= 69.95 - 130.68 - 0.5 * 205.14$$

$$= -163.3\ J\ K^{-1}\ mol^{-1}$$

$$\Delta DG^o_{water} = \Delta H - T\ DS \quad \text{(H in kJ/ mol and S in J/ mol)}$$

$$\Delta DG^o_{water} = -285.83 - 298.15 * 163.3 / 1000 = -237.13\ kJ$$

The equilibrium constant and Gibb's energy are related,

$$\Delta G^o = -R\ T\ \ln K$$

$$K = \exp(-\Delta G^o / R\ T)$$

$$= 3.5 e 41\ atm^{-3/2}$$

This is a very large value for the formation of water,

$$H_2 + 0.5\,O_2 = 0.5\,H_2O$$

In other words, the reaction is complete, and the possibility of water dissociated into hydrogen and oxygen is very small. A negative value for ΔDG^o indicates an exothermic reaction.

The Gibb's energy is the energy released other than pressure-volume work. This redox reaction to form water can be engineered to proceed in a Daniel cell. In this case, the energy is converted into electric energy according to this equation.

$$\Delta G^o_{water} = -n\,F\,E = -237.13 \text{ kJ}$$

where n is the number of electrons (= 2) in the redox equation, F is the Faraday constant (= 96485 C), and E is the potential of the Daniel cell. Thus,

$$E = -\ \frac{-237130\ J}{2*96485\ C}$$

$$= 1.23\ V$$

Ideally, a reverse voltage of 1.23 V is required for the electrolysis of water. But in reality, a little over voltage is required to carry out the electrolysis to decompose water. Furthermore, pure water does not conduct electricity, and acid, base or salt is often added for the electrolysis of water.

Electrolysis of Water

The electrolysis of water is a process that uses an electrical current to split water molecules into hydrogen and oxygen. The process requires three components: an electrical source, two electrodes and water.

Pure water is not used in electrolysis — pure water inhibits electrical conduction. To allow the electrical current to pass through the water, substances must be added to it. These substances dissolve to form something called electrolytes.

An electrolyte is any substance that conducts electricity. Electrolytes are able to conduct electricity because they are composed of electrically charged atoms or molecules called ions. Although water is composed of hydrogen and oxygen ions, the water molecule itself has a neutral electrical charge. Salt or a few drops of an acid or base are commonly added to the water to form an electrolyte solution.

Batteries, a direct current (DC) power source or solar electrical panels are commonly used to supply electricity for the electrolysis of water. Two electrodes are wired to the electrical source and immersed in a container of water. When electricity is applied, the water molecules begin to split, forming unstable ions of hydrogen (H^+) and hydroxide (OH^-).

The hydrogen ions, which are missing an electron, are positively charged. They migrate toward the negative electrode where free electrons are flowing into the water. Here the hydrogen ions gain an electron to form stable hydrogen atoms. Individual hydrogen atoms combine to form hydrogen molecules (H^2), which bubble to the surface. This reaction can be expressed as: $2\,H^+ + 2\,e^- \rightarrow H_2$.

By contrast, the hydroxide ions carry too many electrons. They migrate toward the positive electrode, where the extra electrons are stripped away and drawn into the electrical circuit. This leaves oxygen molecules and water. This reaction can be expressed as: $4\,OH^- - 4\,e^- \rightarrow O_2 + 2\,H_2O$. The oxygen molecules bubble to the surface.

Although the electrolysis of water has been mainly confined to laboratories, the use of hydrogen as a clean energy source has brought renewed interest. Finding a clean energy source to drive the reaction, however, poses practical and environmental concerns. The electrolysis of water is neither efficient nor cheap.

Fuel costs have been a major obstacle. The environmental impact of electrical generation is another. In particular, the carbon dioxide released by thermal power plants must be considered. These environmental and technological difficulties might not prove insurmountable. Until they are overcome, however, water hydrolysis remains an impractical source for fulfilling society's energy needs.

Water of Crystallization

In chemistry, water of crystallization or water of hydration is water molecules that are present inside crystals. Water is often incorporated in the formation of crystals from aqueous solutions. In some contexts, water of crystallization is the total mass of water in a substance at a given temperature and is mostly present in a definite (stoichiometric) ratio. Classically, "water of crystallization" refers to water that is found in the crystalline framework of a metal complex or a salt, which is not directly bonded to the metal cation.

Upon crystallization from water or moist solvents, many compounds incorporate water molecules in their crystalline frameworks. Water of crystallization can generally be removed by heating a sample but the crystalline properties are often lost. For example, in the case of sodium chloride, the dihydrate is unstable at room temperature.

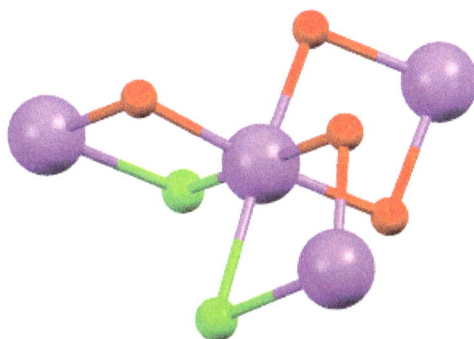

Coordination sphere of Na⁺ in the metastable dihydrate of sodium chloride
(red = oxygen, violet = Na⁺, green = Cl⁻, H atoms omitted).

Compared to inorganic salts, proteins crystallize with large amounts of water in the crystal lattice. A water content of 50% is not uncommon for proteins.

Nomenclature

In molecular formulas water of crystallization can be denoted in different ways:

- "hydrated compound·nH_2O" or "hydrated compound×nH_2O".

  This notation is used when the compound only contains *lattice water* or when the crystal structure is undetermined. For example Calcium chloride: $CaCl_2·2H_2O$.

- "hydrated compound$(H_2O)_n$".

  A hydrate with coordinated water. For example Zinc chloride: $ZnCl_2(H_2O)_4$.

- Both notations can be combined as for example in copper(II) sulfate: $[Cu(H_2O)_4]SO_4·H_2O$.

Position in the Crystal Structure

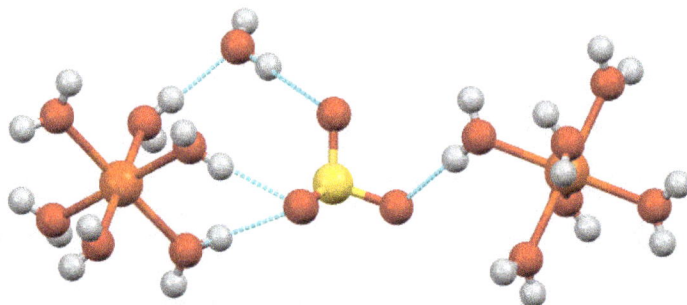

Some hydrogen-bonding contacts in $FeSO_4·7H_2O$. This metal aquo complex crystallizes with water of hydration, which interacts with the sulfate and with the $[Fe(H_2O)_6]^{2+}$ centers.

A salt with associated water of crystallization is known as a hydrate. The structure of hydrates can be quite elaborate, because of the existence of hydrogen bonds that define polymeric structures. Historically, the structures of many hydrates were unknown, and the dot in the formula of a hydrate was employed to specify the composition without indicating how the water is bound. Examples:

- $CuSO_4·5H_2O$ - copper(II) sulfate pentahydrate

- $CoCl_2·6H_2O$ - cobalt(II) chloride hexahydrate

- $SnCl_2·2H_2O$ - tin(II) (*or* stannous) chloride dihydrate

For many salts, the exact bonding of the water is unimportant because the water molecules are labilized upon dissolution. For example, an aqueous solution prepared from $CuSO_4·5H_2O$ and anhydrous $CuSO_4$ behave identically. Therefore, knowledge of the degree of hydration is important only for determining the equivalent weight: one mole of $CuSO_4·5H_2O$ weighs more than one mole of $CuSO_4$. In some cases, the degree of hydration can be critical to the resulting chemical properties. For example, anhydrous $RhCl_3$ is not soluble in water and is relatively useless in organometallic chemistry whereas $RhCl_3·3H_2O$ is versatile. Similarly, hydrated $AlCl_3$ is a poor Lewis acid and thus inactive as a catalyst for Friedel-Crafts reactions. Samples of $AlCl_3$ must therefore be protected from atmospheric moisture to preclude the formation of hydrates.

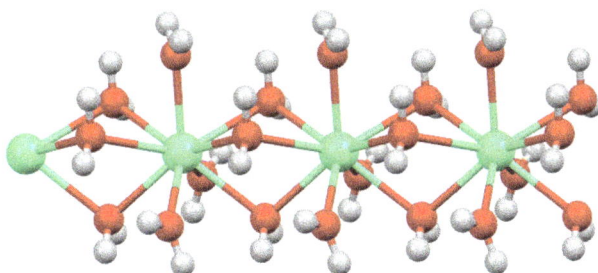

Structure of the polymeric $[Ca(H_2O)_6]^{2+}$ center in crystalline calcium chloride hexahydrate. Three water ligands are terminal, three bridge. Two aspects of metal aquo complexes are illustrated: the high coordination number typical for Ca^{2+} and the role of water as a bridging ligand.

Crystals of hydrated copper(II) sulfate consist of $[Cu(H_2O)_4]^{2+}$ centers linked to SO_4^{2-} ions. Copper is surrounded by six oxygen atoms, provided by two different sulfate groups and four molecules of water. A fifth water resides elsewhere in the framework but does not bind directly to copper. The cobalt chloride mentioned above occurs as $[Co(H_2O)_6]^{2+}$ and Cl^-. In tin chloride, each Sn(II) center is pyramidal (mean O/Cl-Sn-O/Cl angle is 83°) being bound to two chloride ions and one water. The second water in the formula unit is hydrogen-bonded to the chloride and to the coordinated water molecule. Water of crystallization is stabilized by electrostatic attractions, consequently hydrates are common for salts that contain +2 and +3 cations as well as −2 anions. In some cases, the majority of the weight of a compound arises from water. Glauber's salt, $Na_2SO_4(H_2O)_{10}$, is a white crystalline solid with greater than 50% water by weight.

Consider the case of nickel(II) chloride hexahydrate. This species has the formula $NiCl_2(H_2O)_6$. Crystallographic analysis reveals that the solid consists of [*trans*-$NiCl_2(H_2O)_4$] subunits that are hydrogen bonded to each other as well as two additional molecules of H_2O. Thus 1/3 of the water molecules in the crystal are not directly bonded to Ni^{2+}, and these might be termed "water of crystallization".

Analysis

The water content of most compounds can be determined with knowledge of its formula. An unknown sample can be determined through thermogravimetric analysis (TGA) where the sample is heated strongly, and the accurate weight of a sample is plotted against the temperature. The amount of water driven off is then divided by the molar mass of water to obtain the number of molecules of water bound to the salt.

Other Solvents of Crystallization

Water is particularly common solvent to be found in crystals because it is small and polar. But *all* solvents can be found in some host crystals. Water is noteworthy because it is reactive, whereas other solvents such as benzene are considered to be chemically innocuous. Occasionally more than one solvent is found in a crystal, and often the stoichiometry is variable, reflected in the crystallographic concept of "partial occupancy." It is common and conventional for a chemist to "dry" a sample with a combination of vacuum and heat "to constant weight."

For other solvents of crystallization, analysis is conveniently accomplished by dissolving the sample in a deuterated solvent and analyzing the sample for solvent signals by NMR spectroscopy.

Single crystal X-ray crystallography is often able to detect the presence of these solvents of crystallization as well. Other methods may be currently available.

Dealkalization of Water

As a general rule of thumb, dealkalization can be used to treat water in boilers operating at less than 700 psi, with feed water containing less than or equal to 50 ppm alkalinity, and with make-up of 1,000 gallons or more per day. Makeup water is the water added to the boiler to offset water lost due to steam and blow down.

Raw water alkalinity may be reduced using several different methods:

- Reverse osmosis: Membrane filtration has become the popular option for boiler water treatment. With appropriate pretreatment, nearly all carbon dioxide can be eliminated from RO-treated feed water. Reverse osmosis can simultaneously remove up to 98% of all dissolved minerals, greatly reducing alkalinity and blow down- limiting minerals. Boiler cycles may be increased to 50 or more with reverse osmosis.

- Chloride anion dealkalizers: Chloride anion dealkalizers operate similar to ion exchange water softeners, except the filtration vessels contain a Type II strong-base anion resin. Two methods may be used for resin regeneration. The first uses salt and the second use a salt-caustic combination (NaOH). If just salt is used, water hardness should be 10 grains or less (<170 ppm) to prevent the precipitation of $CaCO_3$. If a salt-caustic combination is used, water must be softened prior to being fed to the dealkalizers. In most cases where a dealkalizer is required, water should be softened to prevent boiler scale buildup. Salt-caustic combinations have a higher capacity for alkalinity before regeneration is required.

- Weak acid dealkalization: When the ratio of water hardness to alkalinity is 1 or greater, a weak acid cation (WAC) resin offers significant cost advantages. A WAC resin exchanges hydrogen for the hardness associated with the water's alkalinity. Then degasification is used to remove carbon dioxide. Caustic may be added in low doses to raise the final pH level if desired.

- Split stream dealkalization: Split stream dealkalization utilizes two beds of strong acid cation (SAC) operating in parallel. One bed operates in the sodium form, acting as a cation exchange softener. The other bed operates in the hydrogen form and acts like the cation vessel of a demineralizer. Sulfuric acid is typically used to regenerate this bed. Feed water flow is divided between the two vessels. Water softened by the sodium based vessel contains all of the alkalinity, while the stream from the hydrogen vessel contains no alkalinity. The streams from each vessel are then blended together and degasified to remove carbon dioxide. The controlled ratio at which each stream is blended will determine the final alkalinity level in the effluent.

Dealkalizers

A dealkalizer works similar to a water softener, in that it utilizes ion exchange to remove unwanted ions from a water supply. However, rather than removing calcium and magnesium ions, dealkalization removes carbonate ions, exchanging them for chloride ions. Like water softeners, dealkalizers make use of salt during the regeneration process. Unlike water softeners, a dealkalizer resin must also be treated with an additional caustic solution. This caustic solution boosts pH levels and enhances the resins efficiency.

Self-ionization of water

The self-ionization of water (the process in which water ionizes to hydronium ions and hydroxide ions) occurs to a very limited extent. When two molecules of water collide, there can be a transfer of a hydrogen ion from one molecule to the other. The products are a positively charged hydronium ion and a negatively charged hydroxide ion.

$$H_2O(l) + H_2O(l) \rightleftarrows H_3O^+(aq) + OH^-(aq)$$

We often use the simplified form of the reaction:

$$H_2O(l) \rightleftarrows H^+(aq) + OH^-(aq)$$

The equilibrium constant for the self-ionization of water is referred to as the ion-product for water and is given the symbol K_w.

$$K_w = [H^+][OH^-]$$

The ion-product of water (K_w) is the mathematical product of the concentration of hydrogen ions and hydroxide ions. Note that H_2O is not included in the ion-product expression because it is a pure liquid. The value of Kw is very small, in accordance with a reaction that favors the reactants. At 25°C, the experimentally determined value of Kw in pure water is 1.0×10^{-14}.

$$K_w = [H^+][OH^-] = 1.0 \times 10^{-14}$$

In pure water, the concentrations of hydrogen and hydroxide ions are equal to one another. Pure water or any other aqueous solution in which this ratio holds is said to be neutral. To find the molarity of each ion, the square root of K_w is taken.

$$[H^+] = [OH^-] = 1.0 \times 10^{-7} \text{ M}$$

An acidic solution is a solution in which the concentration of hydrogen ions is greater than the concentration of hydroxide ions. For example, hydrogen chloride ionizes to produce H^+ and Cl^- ions upon dissolving in water.

$$HCl(g) \rightarrow H^+(aq) + Cl^-(aq)$$

This increases the concentration of H^+ ions in the solution. According to LeChâtelier's principle, the equilibrium represented by $H_2O(l) \rightleftarrows H^+(aq) + OH^-(aq)$ is forced to the left, towards the reactant. As a result, the concentration of the hydroxide ion decreases.

A basic solution is a solution in which the concentration of hydroxide ions is greater than the concentration of hydrogen ions. Solid potassium hydroxide dissociates in water to yield potassium ions and hydroxide ions.

$$KOH(s) \rightarrow K^+(aq) + OH^-(aq)$$

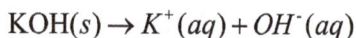

The increase in concentration of the OH^- ions causes a decrease in the concentration of the H^+ ions and the ion-product of $[H^+][OH^-]$ remains constant

Sample Problem: Use of K_w. for an Aqueous Solution

Hydrochloric acid (HCl) is a strong acid, meaning it is 100% ionized in solution. What is the $[H^+]$ and the $[OH^-]$ in a solution of 2.0×10^3 M HCl?

Step 1: List the known values and plan the problem.

- $[HCl] = 2.0 \times 10^{-3} M$
- $K_w = 1.0 \times 10^{-14}$

Unknown

- $[H^+] = ? M$
- $[OH^-] = ? M$

Because HCl is 100% ionized, the concentration of H^+ ions in solution will be equal to the original concentration of HCl. Each HCl molecule that was originally present ionizes into one H+ ion and one Cl^- ion. The concentration of OH^- can then be determined from the $[H^+]$ and K_w.

Step 2: Solve.

$$[H^+] = 2.0 \times 10^{-3} M$$

$$K_w = [H^+][OH^-] = 1.0 \times 10^{-14}$$

$$[OH^-] = \frac{K_w}{[H^+]} = \frac{1.0 \times 10^{-14}}{2.0 \times 10^{-3}} = 5.0 \times 10^{-12} M$$

Step 3: Think about your result.

The $[H^+]$ is much higher than the $[OH^-]$ because the solution is acidic. As with other equilibrium constants, the unit for K_w is customarily omitted.

References

- Moeller, Therald (Jan 1, 1980). Chemistry: With Inorganic qualitative Analysis. Academic Press Inc (London) Ltd. p. 909. ISBN 0-12-503350-8. Retrieved 15 June 2014

- Fresh-water-and-children, global-water-scarcity: eschooltoday.com, Retrieved 30 June 2018

- Klewe, B.; Pedersen, B. (1974). "The crystal structure of sodium chloride dihydrate". Acta Crystallogr. B30: 2363–2371. doi:10.1107/S0567740874007138

- Seawater, science: britannica.com, Retrieved 20 April 2018

- Greenwood, Norman N.; Earnshaw, Alan (1997). Chemistry of the Elements (2nd ed.). Butterworth-Heinemann. ISBN 0-08-037941-9

- Water-cycle, encyclopedia: nationalgeographic.org, Retrieved 10 April 2018

- "Structure Cristalline et Expansion Thermique de L'Iodure de Nickel Hexahydrate" (Crystal structure and thermal expansion of nickel(II) iodide hexahydrate) Louër, Michele; Grandjean, Daniel; Weigel, Dominique Journal of Solid State Chemistry (1973), 7(2), 222-8. doi: 10.1016/0022-4596(73)90157-6

- What-is-the-electrolysis-of-water: wisegeek.com, Retrieved 12 March 2018

- Zalkin, Allan; Forrester, J. D.; Templeton, David H. (1964). "Crystal structure of manganese dichloride tetrahydrate". Inorganic Chemistry. 3: 529–33. doi:10.1021/ic50014a017

- Dealkalization, learning-center: waterprofessionals.com, Retrieved 24 July 2018

5

Natural Hazards

Natural hazards are natural phenomena, which have a detrimental effect on human beings and the environment. These can be broadly classified into geological and meteorological hazards. This chapter discusses in elaborate detail the different typess of natural hazards such as earthquake, coastal erosion, landslide, hurricane, tornado, flood, etc.

Natural hazards are extreme natural events that can cause loss of life, extreme damage to property and disrupt human activities.

Some natural hazards, such as flooding, can happen anywhere in the world. Other natural hazards, such as tornadoes, can only happen in specific areas. And some hazards need climatic or tectonic conditions to occur, for example tropical storms or volcanic eruptions.

Human activities can influence how often certain natural hazards occur and how severe they are. Understanding when, where, why and how natural hazards occur can help us to understand how to minimize their impact on our lives.

The aftermath of Hurricane Katrina

Common Types of Natural Hazards

Natural hazards can be classified into several broad categories: geological hazards, hydrological hazards, meteorological hazards, and biological hazards.

Geological hazards are hazards driven by geological (i.e., Earth) processes, in particular, plate tectonics. This includes earthquakes and volcanic eruptions. In general, geological extreme events are beyond human influence, though humans have a large influence on the impacts of the events.

Meteorological hazards are hazards driven by meteorological (i.e., weather) processes, in particular those related to temperature and wind. This includes heat waves, cold waves, cyclones, hurricanes, and freezing rain. Cyclones are commonly called hurricanes in the Atlantic and typhoons in the pacific ocean.

Hydrological hazards are hazards driven by hydrological (i.e., water) processes. This includes floods, droughts, mudslides, and tsunamis. Floods and droughts can cause extensive damage to agriculture and are among the main contributors to famine. The deadliest natural disaster in world history (not counting pandemics) was the 1931 Central China floods, killing three or four million people.

Biological hazards are hazards driven by biological processes. This includes various types of disease, including infectious diseases that spread from person to person, threatening to infect large portions of the human population. Many discussions of natural hazards exclude biological hazards, placing them instead within the realm of medicine and public health. If biological hazards are counted, then they include the deadliest disasters in world history, including the Black Death outbreak of bubonic plague in the 1300s, killing 75-100 million people, and the 1918 "Spanish" flu pandemic, a global affair (the name "Spanish" is due to historical coincidence) killing 50-100 million people. While biological hazards are undoubtedly important.

It is possible for an extreme event to fit within more than one of these categories. For example, volcano eruptions (a geological event) block incoming sunlight, potentially enough to cause cold waves (a meteorological event). This happened in dramatic fashion in 1816 when the Mount Tambora eruption caused the 'year without summer' in the Northern hemisphere. Volcano eruptions can also cause tsunamis (a hydrological event); some of the largest tsunamis ever occurred when volcanoes along coasts caused large landslides into the water. Earthquakes (a geological event) that occur under water can also trigger tsunamis (a hydrological event), such as the 2011 Japan Earthquake and Tsunami.

Systems of Hazards

One extreme event can often be hazardous in several ways. For instance, an earthquake may destroy buildings, cause landslides, and rupture sewer and water lines. The ruptured lines may, in turn, contaminate water, causing water-borne diseases such as cholera. Indeed, a cholera outbreak happened after the 2010 Haiti earthquake because of disruptions to clean water supplies.

Likewise, a single natural hazard can have many impacts. For instance, hurricanes involve high winds, torrential rain, flooding, and storm surges. The winds may remove roofs and topple power lines. The floods may inundate roads, homes and schools. Ecosystems can be damaged, threatening wildlife. Some impacts can even be beneficial. A hurricane churns up ocean water, cooling surface water and thus reducing the risk of another hurricane in the same area. Keeping track of these systems of hazards and impacts is an important part of the study of hazards.

Hazard Risks

Hazards can have economic, social and environmental consequences. For each hazard event the risks, or probability, of a particular consequence occurring can vary greatly.

This depends on certain factors. For example in a developing country, the death toll tends to be high but the short-term economic costs are often relatively low, whereas in a developed country, the death toll tends to be low but the short-term economic costs can be extremely high.

The long-term situation is more complex. Developing countries can be slower to repair damage to roads and buildings. This can lead to a reduction in tourists and therefore a long-term loss of valuable income.

Hazard risks are increasing due to population growth, urbanisation, pressure on marginal land and changes to the natural environment.

Geological Hazards

Geologic hazards are naturally occurring (or man-made) geologic conditions capable of causing damage or loss of property and/or life. Geologic Hazards Mitigation is the application of geologic engineering principles to minimize or prevent the effects of naturally occurring geologic hazards.

Geologic hazards phenomena can occur suddenly, or slowly. Sudden phenomena include:

- Earthquakes: liquefaction (soils), tsunamis.

- Volcanic eruptions: lava flows, ash fall, lahars.

- Landslides: rock falls or slides, debris flows, mud flows.

- Floods: inundation, erosion.

- Snow avalanches.

- Sand blasting (windblown).

Gradual or slow developing geologic phenomena include:

- Ground settlement.

- Ground subsidence or collapse.

- Sinkholes.

- Erosion (stream or shoreline).

Geologic hazards can play a significant role when infrastructure is constructed in their presence. The unpredictable nature of natural geologic hazards makes identifying, evaluating, and mitigating against them a unique challenge. The best geologic mitigation strategy is always avoidance. When avoiding a hazard is not possible however, mitigation strategies must be developed to coexist (mitigate) with the hazard.

Earthquake

Of all the natural hazards, earthquakes release the most energy in the shortest possible time. On average, each year earthquakes kill 10 000 people and cause US$20 billion property damage. Earthquakes can be regarded as one of the most destructive forces for human beings.

Earthquakes demonstrate that the Earth continues to be a dynamic planet, changing each day through internal tectonic forces. The crust of the Earth consists of various elastic rocks in which energy is stored during crustal deformation caused by the tectonic forces. When the strain builds to a level that exceeds the strength of a weak part of the Earth's crust, such as along a geological fault, then opposite sides of the fault suddenly slip, and an earthquake occurs. The common parameters for describing the characteristics of an earthquake source are the location of the hypocenter or the epicenter (the point on the Earth's surface immediately above the hypocenter). Measures of the strength of shaking and the total energy release in the earthquake are also needed.

We know that the Earth's crust is not a continuous skin; instead it is like a completed jigsaw puzzle with the actual pieces of crust termed "plates." Most earthquakes occur along the plate boundaries, which are called inter-plate earthquakes, other earthquakes occur in the inner parts of continents; these are called intra-plate earthquakes. The intra- plate earthquakes are more dangerous to human beings because most people live in continental regions.

In most cases, the empirical relation between magnitude m and seismic wave energy released E (unit: ergs) can be written as:

$$\log E(\text{in erg}) = 11.8 + 1.5m$$

This equation indicates an about thirty-fold $(10^{1.5})$ increase in seismic wave energy when the magnitude m increases by one unit. For example, the seismic energy released by an earthquake of magnitude m=6.5 is about 30 times greater than that of an event of magnitude $m = 5.5$ (which is the same as that released by the explosion of the atomic bomb in Hiroshima in 1945), and the seismic energy release of an event of $m = 7.5$ is about $30 \times 30 \approx 1000$ times greater than that of $m = 5.5$ (equivalent to about 1000 Hiroshima atomic bombs).

Earthquakes are generally regarded as the most destructive of all the various natural forces. The comparison of energy released by earthquakes and other kinds of energy in nature.

Modern seismographic networks record millions of earthquakes every year; over 99% of these events pose no danger because they are small. An important scaling relationship is the relation between earthquake size and frequency of occurrence. Gutenberg and Richter first proposed that in a given region and over a given period of time, the frequency of occurrence could be represented by:

$$log\ N(\geq m) = A - b\ m$$

Where the N $N(\geq m)$ is the number of earthquakes with magnitude m or above, A and b are empirical constants determined through statistical study, and m is the magnitude of earthquakes.

Description	Magnitude	Average annual frequency	Energy released (ergs)
Great	>8.0	1	$> 5.8 \times 10^{23}$
Major	7.0~7.9	18	$2 \sim 42 \times 10^{22}$
Strong	6.0~6.9	120	$8 \sim 150 \times 10^{20}$
Moderate	5.0~5.9	800	$3 \sim 55 \times 10^{20}$
Light	4.0~4.9	6200	$1 \sim 20 \times 10^{20}$
Minor	3.0~3.9	49 000	$1 \sim 26 \times 10^{20}$

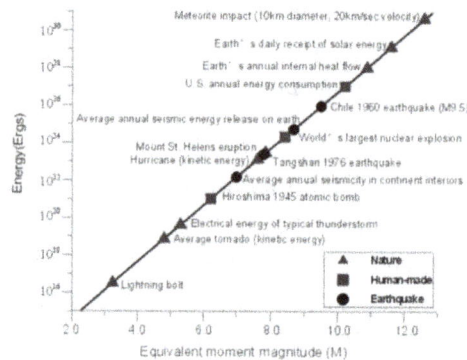

Earthquake Hazards

Strangely, the release of all energy from earthquakes beneath the surface of the Earth poses little direct danger to the individual person. Humans are not "shaken to death" by earthquakes. The greatest danger comes from the interaction between the ground motion caused by earthquakes and man's own structures. The dangers of being crushed in a falling building, getting burned by fire, being swept away and drowned in a flood from a burst reservoir, or getting buried beneath earthquake-induced landslides are very real.

Earthquake-caused damages include the following four aspects:

- Ground shaking is generally the most severe direct cause of damage. Crowded buildings that cannot be evacuated quickly may collapse during ground-shaking and result in a major loss of life as well as property.

- Surface rupture is the horizontal or vertical displacement of the ground surface along the narrow fault zone. While affecting a much smaller area compared to ground shaking, it can severely damage structures located adjacent to faults.

- Ground failure is an indirect cause of damage, but it may be widespread and produce some of the most devastating loss of life.

- Tsunamis are ocean waves produced by earthquakes, which may sweep ashore, causing damage at points thousands of kilometers from the earthquake epicenter.

Damage can be severe where the waves move forward up the shoreline or over dams, allowing downstream areas to be inundated.

Ground Shaking

Traditionally, engineers have been interested in acceleration, particularly Peak Ground Acceleration (PGA), which is related to the dynamic force and can be reliably measured. The unit of PGA is g, which is the gravitational acceleration at Earth's surface (1 g is approximately 9.8 m/sec2).

The most powerful vibrations from an earthquake are in frequency range 0.5–5 Hz and are at near and regional distance. A typical building of ten stories has a natural period of about 1 s. Each story adds about 0.1 s and a 20-storey building has a period near 2 s. Taller buildings have the advantage of flexing more than short, stiff buildings, and is usually designed to bend with the wind. Thus, those over 20 stories may fare relatively well. But, with their longer periods, they are sensitive to distant earthquakes, such as that in Mexico City in 1985. Therefore, anti-seismic design may include estimates of the buildings' responses and the frequencies of various vibration modes.

It should be emphasized that multi-parameters for describing ground motion are really needed because the effect of ground motion on the damage of buildings depends upon the amplitudes, duration, and frequency content of ground motion. In addition to the above mentioned ground motion parameter PGA, there is another form of ground motion effect—intensity scale created first by M. S. de Rossi of Italy and F. A. Forel of Switzerland at the end of the nineteenth century. The Rossi-Forel Intensity Scale (RF), has proved to be of great importance in the evaluation of ground motion from the point of view of earthquake hazard. Often the choice of earthquake intensity scale is a matter of local, that is, national preference. For example, in Southern Europe, the 12-level Mercalli-Cancani-Sieberg intensity scale (MCS) is used. The intensity scale used by the Japanese Meteorological Agency (JMA) is based on seven levels that refer to earthquake effects on typical Japanese items such as latticed sliding doors and wooden houses. Alternatively, the Medvedev-Sponheuer-Karnik intensity scale (KSK), a 12- level scale, was developed by Central and Eastern European scientists. The intensity scale currently used in China and United State is the Modified Mercalli Intensity scale (MMI). An increase in intensity describes a more severe effect on what people feel and what can be observed around them. Figure gives a comparison of various seismic intensity scales.

A comparison of different seismic intensity scales

Surface Rupture and Other Related Hazards

Surface rupture and movement along the fault are obvious hazards. The offset between rocks on the surface rupture, or on the opposite sides of the fault, can break power lines, pipelines, buildings, roads, bridges, and other structures that actually cross the fault.

Surface rupture sometimes closely relates to shaking. Ground shaking may cause a further problem in areas where it is relatively wet. The process by which poorly consolidated mud and other fine-grained sediments become fluid during shaking is called "liquefaction," and it affects what appear to be solid, compact mud or silts. When wet soil is shaken by an earthquake, the soil particles may be jarred apart, allowing water to seep in between them, reducing the friction between soil particles that gives the soil strength, and causing the ground to become somewhat like quicksand. When this happens, buildings can just topple over or partially sink into the liquefied soil; the soil has no strength to support them.

Solid bedrock is the most stable foundation, and buildings on it have a good chance of riding out all but the most severe earthquakes. Where the underlying soils or sediments are weak and poorly consolidated, however, the story is different. The risk factor from ground displacement is often exacerbated in urban areas where land is at a premium, and many cities have expanded into wetlands and shallow coastal regions by using artificial fill to increase the land area.

Indirect Hazards

An indirect hazard of earthquakes in cities is fire, which may be more devastating than ground movement. Prior to the modern electrical service, most city dwellers used wood- or coal-burning stoves for heat and cooking and open flames or lanterns for light, all of which were often toppled by the shaking during earthquakes. Even today, the combination of electrical short-circuits caused by destruction of service poles and transformers and the presence of broken gas mains can produce enormous risk from fire. In Kobe, about 10% of the fatalities were fire-related, about two-thirds apparently caused by leaking gas or electrical problems. The problem of fire is exacerbated by broken water mains, loss of water pressure, and the inability of fire companies to negotiate the rubble-strewn streets of an earthquake-damaged city.

As with fire, the collapse of the social infrastructure—municipal water supplies, sewage treatment facilities, burial of the dead, isolation of outlying areas from food and medical care—contributes to a general decline of social services. Cholera and other epidemics are common in more remote areas of the less-developed world in the aftermath of earthquake destruction. In 1993, an earthquake centered on Khillari, 300 miles southeast of Bombay, killed perhaps as many as 22 000 people. In the aftermath, shortages of water and proper sanitation resulted in epidemics of gastro-enteritis and malaria, although the far more dangerous spread of cholera and diphtheria was prevented by the rapid response of public health officials.

Earthquake Hazard Assessment

In earthquake hazards assessment, the first question we should make clear is where the danger is and, therefore, who is in danger and to what degree the loss will be. The answer to these questions is critical to earthquake disaster mitigation and can be broken down into two main aspects: seismic hazard analysis, which involves the identification and quantitative description of strong ground motion caused by future earthquakes; and seismic risk analysis, which involves the vulnerability analysis of buildings and other man-made facilities to earthquake damage, and the losses that may result from this damage.

Earthquake hazard is the probability that a certain value of a macroscopic intensity or of a ground motion parameter (i.e. particle acceleration, velocity and displacement) will not be exceeded at a specific site in a specific period of time.

Today, many maps have been developed to help public officials prepare for earthquakes. Such maps are based on mapping active faults, studies of geologic features that allow dating of earthquake-produced scarps, landslides, offsets and liquefaction features, and the historical record of seismic activity. Figure is a global seismic hazard map compiled the Global Seismic Hazard Assessment Program (GSHAP). GSHAP was launched in 1992 by the International Lithosphere Program (ILP) with the support of the International Council of Scientific Unions (ICSU), and endorsed as a demonstration program in the framework of the United Nations International Decade for Natural Disaster Reduction (UN/IDNDR). The GSHAP project terminated in 1999.

GLOBAL SEISMIC HAZARD MAP

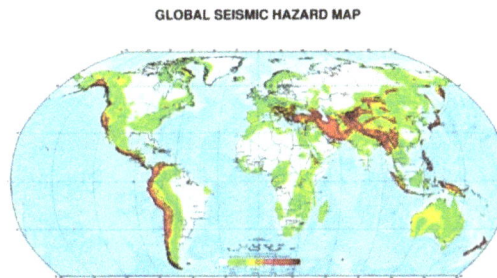

Seismic risk is the expected degree of losses caused by earthquakes and therefore the product of seismic hazard and vulnerability. Vulnerability is the expected degree of loss within a defined area resulting from the occurrence of earthquakes. Vulnerability is expressed on a scale of zero (no damage) to one (full damage). Thus, an equation could be used like this:

$$\text{Risk} = \text{Hazard} \times \text{Vulnerability}$$

Seismic hazard describes the potential for dangerous, earthquake-related natural phenomena such as ground shaking, fault rupture, or soil liquefaction. These phenomena could result in adverse consequences to society, such as the destruction of buildings or the loss of life. Seismic risk is the probability of occurrence of these consequences. The output of a seismic hazard analysis could be a description of the intensity of shaking of a nearby magnitude eight earthquake or a map which shows levels of ground shaking in various parts of the country that have an equal chance of being exceeded.

The results of seismic hazard analysis provide basis for anti-earthquake design, which is the main engineering measure for the reduction of earthquake disasters. The seismologists performing seismic hazard analysis are really carrying out one part of an engineering process or one part of the social disaster reduction process. The end-product of this analysis is an expression of seismic hazard or threat that is oriented toward some specific use. This product may be in the form of simple, single-value characterizations of earthquake ground motion such as Modified Mercalli Intensity (MMI) or Peak Ground Acceleration (PGA), or the more complex, multi-value characterizations, such as response spectra.

Coastal Erosion

Natural coastal erosion of sandy coasts is caused by many factors. Coastal erosion may also result from man-induced activities. Erosion due to human interventions is described in e.g. Human causes of coastal erosion.

Temporal Erosion and Ongoing Erosion

An important notion for understanding coastal erosion refers to the concept of coastal sediment cell. A coastal sediment cell (also called littoral cell or sediment cell) is a coastal compartment that contains a complete cycle of sedimentation including sources, transport paths, and sinks. Erosion at one place in a coastal sediment cell implies accretion at another location within the same cell; the sediment distribution within the cell changes without affecting other coastal regions. A coastal sediment cell is in morphodynamic equilibrium if changes in the sediment distribution under the influence of fluctuating forcing (fluctuations in water levels, wave climate, including storms) have a temporal, quasi-cyclic character. Ongoing trends of erosion or accretion are excluded; ongoing erosion or accretion will finally lead to destruction or basic alteration of the coastal sediment cell. In practice there is always a net leakage of sediment from or to other coastal regions, but this can be a very slow process. So even though the coastal sediment cell is a theoretical concept, it can be very useful in practice for analyzing and managing coastal erosion processes.

One example of a coastal sediment cell is a pocket beach enclosed between headlands, assuming absence of net offshore or onshore sand transport. The orientation of the beach can change in response to fluctuations in the dominant direction of incident waves. However, the resulting erosion and accretion have just a temporal character.

Another example of a coastal sediment cell refers to the concept of active coastal zone. The active coastal zone (sometimes also called active coastal profile) is the beach zone over which sand is exchanged in cross-shore direction by natural processes. The seaward limit corresponds to the closure depth and the landward limit to a hard boundary (seawall, cliff). In the case of a dune coast the active zone comprises part of the front dune that can be eroded by storm waves. In the absence of net offshore or onshore sand transport and in the absence of gradients in littoral drift the active coastal zone is a one-dimensional sediment cell; the sediment volume in the active zone will be constant in time. Shoreline erosion and accretion in response to fluctuations in water level and wave climate (including storms) are temporal quasi-cyclic phenomena in this case.

An obvious example of ongoing erosion is cliff erosion. A coastal cliff eroded by wave attack will never be rebuilt by natural processes. Cliff erosion can be a fast process in case of soft cliffs (till, clay) and very slow in case of hard-rock cliffs. Other examples of ongoing natural erosion are given below.

Long Shore Sand Loss Due to Transport Gradients

One cause of ongoing natural coastal erosion is a long shore increase of sand transport: in this case more sand is leaving a coastal section than entering. As long shore sand transport (also called littoral drift) depends primarily on the direction and height of breaking waves, a gradient in long shore transport can be due to long shore varying wave conditions, coastline curvature, or near shore bathymetric features. An example of this kind of coastal condition is the West Coast of Skaw Spit, the northernmost tip of Denmark. The presence of the headland and the port at Hirtshals, combined with the shadow effect of southern Norway, results in increasing transport along the section of coastline some kilometres east of Hirtshals to Gammel Skagen. For this reason the entire NW-oriented section of the Skaw spit is exposed to erosion.

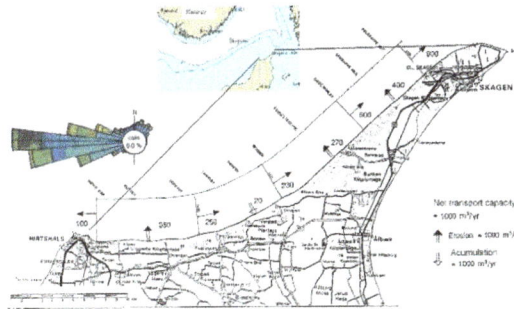

Cross-shore Sand Loss

Sand loss from the active coastal zone in cross-shore direction can occur by different processes.

Breaching and Over-wash

The loss of sand inland due to breaching and over-wash of a barrier island. This kind of sand loss takes place along the exposed coast of Skallingen barrier island on the southern part of the Danish North Sea coast, see figure above.

Aeolian Transport to the Dunes

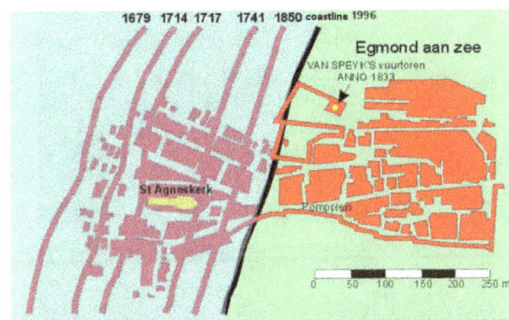

A wide vegetated dune area can trap fine sands carried inland from the beach by onshore winds. When the dunes along the Dutch coast were fixed by vegetation from the 16th century they started growing by capturing large amounts of sand. This contributed to fast shoreline retreat, illustrated in figure above. The import of beach sand to the dune area at the Dutch is estimated at 5-35 m3/m/year Assuming that this sand volume is withdrawn from the active zone (width of the order of 2 km, average slope of the order of 1/100), the resulting shoreline retreat can be estimated at 0.2-2 m/year.

Offshore Sand Loss Under Extreme Wave and Storm Surge Conditions

High energetic waves cause seaward migration of breaker bars and high storm surges further cause an offshore movement of sand due to non-equilibrium in the profile during the high surge. Sand that is transported sufficiently far offshore will not return to the coast by wave-induced onshore transport under a milder wave climate.

Offshore Sand Loss to Canyons

If there is a deep canyon close to the shore, sand may be lost into the canyon by littoral drift.

Offshore Transport at the Tip of a Sand Spit

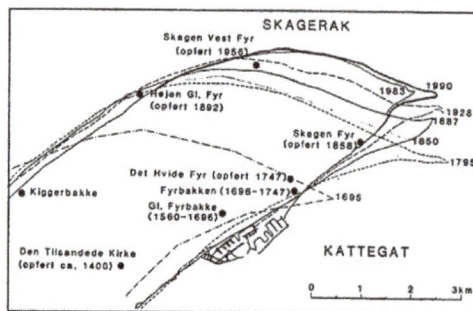

A similar process of natural coastal erosion occurs downstream of accumulative forms at coastlines with very oblique wave approach, coast types 4M, 4E, 5M and 5E. Along such coastlines there is a tendency for the natural formation of spits parallel to the coast. They accumulate the sand and shift the sand supply offshore, which means that the downstream coastline is starved and begins to erode, see figure below.

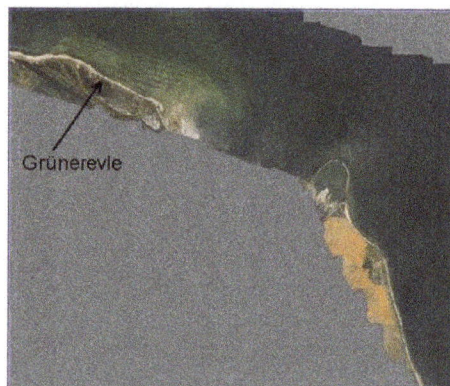

Sand Loss at Coastal Protrusions

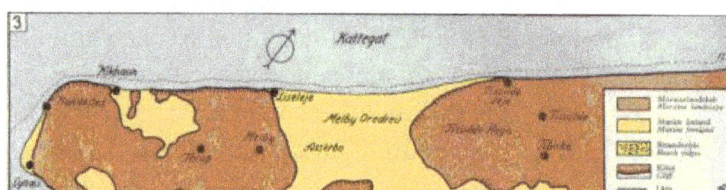

The loss of material from a protruding area to one or two sides is a natural cause of coastal erosion. This typically happens at till/sandstone headlands, where fine eroded material is washed away by currents and coarse material is transported alongshore or offshore away from the headland. More generally, any semi-hard seaward-concave section of a coastline will suffer erosion in case of insufficient supply of sand from rivers. The natural state of such a coastline is erosion and straightening; the straightened coastline is referred to as a simplificated coast.

Marine deposit shorelines suspended between eroding headlands (till or sandstone, for example) will retreat similarly. The headlands have historically provided material for building up the sedimentary shorelines and the suspended shoreline is consequently dependent on the presence of the headlands. However, as the headlands continue to erode, the sedimentary shorelines will follow suit despite the fact that they were originally accumulative forms. This development is part of the simplificated coast.

Erosion also occurs at deltas coasts when the natural fluvial sand supply is reduced, depriving the delta protrusion from fluvial sand supply for its maintenance. This is generally caused by human interventions, but it can also have natural causes. Droughts in large river basins can result in long periods with almost no sand supply to the shoreline, leading to shore erosion. The historic large variations in the shorelines of the Nile delta were partly due to this situation, whereas the more recent erosion is mainly the result of human interventions along the Nile. Natural shifts between distributary channels are another cause of delta erosion.

Climate Change Impacts

Climate change will impact on coastal erosion in different ways. Here the focus is on sea-level rise; other potential impacts are related to changes in meteorological conditions – wind, temperature and precipitation. Changes in the precipitation regime will affect the sediment discharge of rivers and the resulting sand supply to the coast. Extreme conditions of strong precipitation and long periods of drought are expected to become more frequent. Temperature may play a role too, by its impact on soil erosion. Change in temperature will affect all life forms in the coastal zone. Coastal erosion is particularly sensitive to changes in coastal vegetation, dune vegetation for example. Mangrove coasts are sensitive to temperature change, but also to sea-level rise. Change in the wind regime and wave climate will modify the alongshore and cross-shore sand distribution. The alongshore sand distribution is very sensitive to the littoral drift, which strongly depends on wave direction. The shape of the cross-shore coastal profile is strongly influenced by wave run-up, with an important role for storm events with high waves and water levels. Great uncertainty still exists regarding predictions for local changes in wind regime and wave climate caused by climate change.

Relative Sea-level Rise

The sea level will rise globally as a consequence of global warming, but regional differences are considerable. This holds in particular for relative sea-level rise, i.e. the change of sea level with respect to the local land level. Some coasts experience uplift (especially in previously glaciated regions) while others are subject to subsidence. Uplift can always be considered "natural", whereas subsidence often has an important human-induced component (groundwater, oil, gas extraction). According to the so-called "Bruun rule", an increasing relative sea level will cause a shoreline setback, which is approximately equal to the sea level rise divided by the average slope of the active coastal profile, when considering equilibrium profiles. Consider, for example, a sea level rise of 0.5 m and an equilibrium coastal profile with a slope of the shore face and the shore of 1/100. The setback caused by such a sea level rise will be 50 m. Littoral coasts consisting of fine sediments will be exposed to higher setbacks than coasts consisting of coarser sediments.

Landslide

A landslide, sometimes known as landslip, slope failure or slump, is an uncontrollable downhill flow of rock, earth, debris or the combination of the three. Landslides stem from the failure of materials making up the hill slopes and are beefed up by the force of gravity. When the ground becomes saturated, it can become unstable, losing its equilibrium in the long run. That's when a landslide breaks loose. When people are living down these hills or mountains, it's usually just a matter of time before disaster happens.

Causes of Landslides

While landslides are considered naturally occurring disasters, human-induced changes in the environment have recently caused their upsurge. Although the causes of landslides are wide ranging, they have 2 aspects in common; they are driven by forces of gravity and result from failure of soil and rock materials that constitute the hill slope.

Natural Causes of Landslides

Climate

Long-term climatic changes can significantly impact soil stability. A general reduction in precipitation leads to lowering of water table and reduction in overall weight of soil mass, reduced solution of materials and less powerful freeze-thaw activity. A significant upsurge in precipitation or ground saturation would dramatically increase the level of ground water. When sloped areas are completely saturated with water, landslides can occur. If there is absence of mechanical root support, the soils start to run off.

Earthquakes

Seismic activities have, for a long time, contributed to landslides across the globe. Any moment tectonic plates move, the soil covering them also moves along. When earthquakes strike areas with steep slopes, on numerous occasion, the soil slips leading to landslides. In addition, ashen debris flows instigated by earthquakes could also cause mass soil movement.

Weathering

Weathering is the natural procedure of rock deterioration that leads to weak, landslide-susceptive materials. Weathering is brought about by the chemical action of water, air, plants and bacteria. When the rocks are weak enough, they slip away causing landslides.

Erosion

Erosion caused by sporadic running water such as streams, rivers, wind, currents, ice and waves wipes out latent and lateral slope support enabling landslides to occur easily.

Volcanoes

Volcanic eruptions can trigger landslides. If an eruption occurs in a wet condition, the soil will start to move downhill instigating a landslide. Stratovolcano is a typical example of volcano responsible for most landslides across the globe.

Forest Fires

Forest fires instigate soil erosion and bring about floods, which might lead to landslides.

Gravity

Steeper slopes coupled with gravitational force can trigger a massive landslide.

Human Causes of Landslides

Mining

Mining activities that utilize blasting techniques contribute mightily to landslides. Vibrations emanating from the blasts can weaken soils in other areas susceptible to landslides. The weakening of soil means a landslide can occur anytime.

Clear Cutting

Clear cutting is a technique of timber harvesting that eliminates all old trees from the area. This technique is dangerous since it decimates the existing mechanical root structure of the area.

Effects of Landslides

Lead to Economic Decline

Landslides have been verified to result in destruction of property. If the landslide is significant, it could drain the economy of the region or country. After a landslide, the area affected normally undergoes rehabilitation. This rehabilitation involves massive capital outlay. For example, the 1983 landslide at Utah in the United States resulted in rehabilitation cost of about $500 million. The annual loss as a result of landslides in U.S. stands at an estimated $1.5 billion.

Decimation of Infrastructure

The force flow of mud, debris, and rocks as a result of a landslide can cause serious damage to property. Infrastructure such as roads, railways, leisure destinations, buildings and communication systems can be decimated by a single landslide.

Loss of Life

Communities living at the foot of hills and mountains are at a greater risk of death by landslides. A substantial landslide carries along huge rocks, heavy debris and heavy soil with it. This kind of landslide has the capacity to kills lots of people on impact. For instance, Landslides in the UK that happened a few years ago caused rotation of debris that destroyed a school and killed over 144 people including 116 school children aged between 7 and 10 years. In a separate event, NBC News reported a death toll of 21 people in the March 22, 2014, landslide in Oso, Washington.

Affects Beauty of Landscapes

The erosion left behind by landslides leaves behind rugged landscapes that are unsightly. The pile of soil, rock and debris downhill can cover land utilized by the community for agricultural or social purposes.

Impacts River Ecosystems

The soil, debris, and rock sliding downhill can find way into rivers and block their natural flow. Many river habitats like fish can die due to interference of natural flow of water. Communities depending on the river water for household activities and irrigation will suffer if flow of water is blocked.

Types of Landslides

Falls

Falls are sudden movements of loads of soil, debris, and rock that break away from slopes and cliffs. Falls landslides occur as a result of mechanical weathering, earthquakes, and force of gravity.

Slides

This is a kind of mass movement whereby the sliding material breakaways from underlying stable material. The kinds of slides experienced during this type of landslide include rotational and transitional. Rotational slides are sometimes known as slumps since they move with rotation.

Transitional slides consist of a planer or 2 dimensional surface of rupture. They involve landslide mass movement following a roughly planar surface with reduced rotation or backward slanting. Slides occur when the toe of the slope is undercut. They move moderately, and the consistency of material is maintained.

Topples

Topple landslides occur when the topple fails. Topple failure encompasses the forward spinning and movement of huge masses of rock, debris, and earth from a slope. This type of slope failure takes place around an axis near or at the bottom of the block of rock. A topple landslide mostly lead to formation of a debris cone below the slope. This pile of debris is known as a Talus cone.

Spreads

They are commonly known as lateral spreads and takes place on gentle terrains via lateral extension followed by tensile fractures.

Flows

This type of landslide is categorized into five; earth flows, debris avalanche, debris flow, mudflows, and creep, which include seasonal, continuous and progressive.

Flows are further subcategorized depending upon the geological material, for example, earth, debris, and bedrock.

The most prevalent occurring landslides are rock falls and debris flow.

Volcanic Eruption

The most common type of volcanic eruption occurs when magma (the term for lava when it is below the Earth's surface) is released from a volcanic vent. Eruptions can be effusive, where lava flows like a thick, sticky liquid, or explosive, where fragmented lava explodes out of a vent. In explosive eruptions, the fragmented rock may be accompanied by ash and gases; in effusive eruptions, degassing is common but ash is usually not.

The fundamental concept of an eruption is that an increase in pressure on the chamber lid causes the magma to be released from beneath it. However, there are variances in the cause of this magma movement and the type of eruption generated.

Volcanoes are usually found near the boundaries of Earth's tectonic plates. These can either spread apart or leave a gap in the surface, or they can push underneath one another – a process called subduction.

When the plates separate, magma rises slowly in order to fill the gap through a gentle explosion of thin basaltic lava, which is at temperatures from 800 to 1200 degrees Celsius.

However, when one plate pushes underneath the other, this forces molten rock, sediment and seawater down into the magma chamber. The rock and sediment are melted into fresh magma, and eventually overfill the chamber until it erupts, releasing sticky and thick andesitic lava, at temperatures from 800 to 1000 degrees Celsius.

Plate tectonics is, however, not the only cause of eruptions.

Decreasing temperatures can cause old magma to crystalize and sink to the bottom of the chamber, forcing fresh liquefied magma up and out – similar to what happens when a brick is dropped in a bucket of water.

A decrease in external pressure on the magma chamber may also allow for an eruption by minimizing its ability to hold back increasing pressures from the inside. This is often caused by natural events, such as typhoons, that decrease rock density, or by glacial melting on top of the chamber lid, which alters molten rock composition. Glacial melting is believed to have been one of the causes of the 2010 Eyjafjallajokull eruption in Iceland.

So-called 'hot-spot' volcanoes are ones that form away from tectonic plate boundaries. They are created as plates move and expose hot uprisings from Earth's mantle, known as plumes. The volcanoes found in the Hawaiian Islands are of this sort.

Types of Volcanic Eruption

Here are some of the most common types of eruptions:

Hawaiian Eruption

In a Hawaiian eruption, fluid basaltic lava is thrown into the air in jets from a vent or line of vents (a fissure) at the summit or on the flank of a volcano. The jets can last for hours or even days, a phenomenon known as fire fountaining. The spatter created by bits of hot lava falling out of the fountain can melt together and form lava flows, or build hills called spatter cones. Lava flows may also come from vents at the same time as fountaining occurs, or during periods where fountaining has paused. Because these flows are very fluid, they can travel miles from their source before they cool and harden.

Hawaiian eruptions get their names from the Kilauea Volcano on the Big Island of Hawaii, which is famous for producing spectacular fire fountains. Two excellent examples of these are the 1969-1974 Mauna Ulu eruption on the volcano's flank, and the 1959 eruption of the Kilauea Iki Crater at the summit of Kilauea. In both of these eruptions, lava fountains reached heights of well over a thousand feet.

Strombolian Eruption

Strombolian eruptions are distinct bursts of fluid lava (usually basalt or basaltic andesite) from the mouth of a magma-filled summit conduit. The explosions usually occur every few minutes at regular or irregular intervals. The explosions of lava, which can reach heights of hundreds of meters, are caused by the bursting of large bubbles of gas, which travel upward in the magma-filled conduit until they reach the open air.

This kind of eruption can create a variety of forms of eruptive products: spatter, or hardened globs of glassy lava; scoria, which are hardened chunks of bubbly lava; lava bombs, or chunks of lava a few cm to a few m in size; ash; and small lava flows (which form when hot spatter melts together and flows downslope). Products of an explosive eruption are often collectively called tephra.

Strombolian eruptions are often associated with small lava lakes, which can build up in the conduits of volcanoes. They are one of the least violent of the explosive eruptions, although they can still be very dangerous if bombs or lava flows reach inhabited areas. Strombolian eruptions are named for the volcano that makes up the Italian island of Stromboli, which has several erupting summit vents. These eruptions are particularly spectacular at night, when the lava glows brightly.

Vulcanian Eruption

A Vulcanian eruption is a short, violent, relatively small explosion of viscous magma (usually andesite, dacite, or rhyolite). This type of eruption results from the fragmentation and explosion of a plug of lava in a volcanic conduit, or from the rupture of a lava dome (viscous lava that piles up over a vent). Vulcanian eruptions create powerful explosions in which material can travel faster than 350 meters per second (800 mph) and rise several kilometers into the air. They produce tephra, ash clouds, and pyroclastic density currents (clouds of hot ash, gas and rock that flow almost like fluids).

Vulcanian eruptions may be repetitive and go on for days, months, or years or they may precede even larger explosive eruptions. They are named for the Italian island of Vulcano, where a small volcano that experienced this type of explosive eruption was thought to be the vent above the forge of the Roman smith god Vulcan.

Plinian Eruption

The largest and most violent of all the types of volcanic eruptions are Plinian eruptions. They are caused by the fragmentation of gassy magma, and are usually associated with very viscous magmas (dacite and rhyolite). They release enormous amounts of energy and create eruption columns of gas and ash that can rise up to 50 km (35 miles) high at speeds of hundreds of meters per second. Ash from an eruption column can drift or be blown hundreds or thousands of miles away from the volcano. The eruption columns are usually shaped like a mushroom (similar to a nuclear explosion) or an Italian pine tree; Pliny the Younger, a Roman historian, made the comparison while viewing the 79 AD eruption of Mount Vesuvius, and Plinian eruptions are named for him.

Plinian eruptions are extremely destructive, and can even obliterate the entire top of a mountain, as occurred at Mount St. Helens in 1980. They can produce falls of ash, scoria and lava bombs miles from the volcano, and pyroclastic density currents that raze forests, strip soil from bedrock and obliterate anything in their paths. These eruptions are often climactic, and a volcano with a magma chamber emptied by a large Plinian eruption may subsequently enter a period of inactivity.

Lava Domes

Lava domes form when very viscous, rubbly lava (usually andesite, dacite or rhyolite) is squeezed out of a vent without exploding. The lava piles up into a dome, which may grow by inflating from the inside or by squeezing out lobes of lava (something like toothpaste coming out of a tube). These lava lobes can be short and blobby, long and thin, or even form spikes that rise tens of meters into the air before they fall over. Lava domes may be rounded, pancake-shaped, or irregular piles of rock, depending on the type of lava they form from.

Lava domes are not just passive piles of rock; they can sometimes collapse and form pyroclastic density currents, extrude lava flows, or experience small and large explosive eruptions (which may even destroy the domes.) A dome-building eruption may go on for months or years, but they are usually repetitive (meaning that a volcano will build and destroy several domes before the eruption ceases). Redoubt volcano in Alaska and Chaiten in Chile are currently active examples of this type of eruption, and Mount St. Helens in the state of Washington spent several years building several lava domes.

Surtseyan Eruption

Surtseyan eruptions are a kind of hydro magmatic eruption, where magma or lava interacts explosively with water. In most cases, Surtseyan eruptions occur when an undersea volcano has finally grown large enough to break the water's surface; because water expands when it turns to steam, water that comes into contact with hot lava explodes and creates plumes of ash, steam and scoria. Lavas created by a Surtseyan eruption tend to be basalt, since most oceanic volcanoes are basaltic.

The classic example of a Surtseyan eruption was the volcanic island of Surtsey, which erupted off the south coast of Iceland between 1963 and 1965. Hydromagmatic activity built up several square kilometers of tephra over the first several months of the eruption; eventually, seawater could no longer reach the vent, and the eruption transitioned to Hawaiian and Strombolian styles. More recently, in March 2009, several vents of the volcanic island of Hunga Ha'apai near Tonga began to erupt. The onshore and offshore explosions created plumes of ash and steam that rose to more than 8 km (5 miles) altitude, and threw plumes of tephra hundreds of meters from the vents.

Meteorological Hazard

Meteorological hazards are hazards driven by meteorological (i.e., weather) processes, in particular those related to temperature and wind. This includes heat waves, cold waves, cyclones, hurricanes, and freezing rain. Cyclones are commonly called hurricanes in the Atlantic and typhoons in the Pacific Ocean.

Hurricane

Hurricanes are large, swirling storms. They produce winds of 119 kilometers per hour (74 mph) or higher. That's faster than a cheetah, the fastest animal on land. Winds from a hurricane can damage buildings and trees.

Hurricanes form over warm ocean waters. Sometimes they strike land. When a hurricane reaches land, it pushes a wall of ocean water ashore. This wall of water is called a storm surge. Heavy rain and storm surge from a hurricane can cause flooding.

Once a hurricane forms, weather forecasters predict its path. They also predict how strong it will get. This information helps people get ready for the storm.

There are five types, or categories, of hurricanes. The scale of categories is called the Saffir-Simpson Hurricane Scale. The categories are based on wind speed.

- Category 1: Winds 119-153 km/hr (74-95 mph) - faster than a cheetah.

- Category 2: Winds 154-177 km/hr (96-110 mph) - as fast or faster than a baseball pitcher's fastball.

- Category 3: Winds 178-208 km/hr (111-129 mph) - similar, or close, to the serving speed of many professional tennis players.

- Category 4: Winds 209-251 km/hr (130-156 mph) - faster than the world's fastest roller-coaster.

- Category 5: Winds more than 252 km/hr (157 mph) - similar, or close, to the speed of some high-speed trains.

Parts of a Hurricane

- Eye: The eye is the "hole" at the center of the storm. Winds are light in this area. Skies are partly cloudy, and sometimes even clear.

- Eye wall: The eye wall is a ring of thunderstorms. These storms swirl around the eye. The wall is where winds are strongest and rain is heaviest.

- Rain bands: Bands of clouds and rain go far out from a hurricane's eye wall. These bands stretch for hundreds of miles. They contain thunderstorms and sometimes tornadoes.

Formation of Hurricane

- A hurricane starts out as a tropical disturbance. This is an area over warm ocean waters where rain clouds are building.

- A tropical disturbance sometimes grows into a tropical depression. This is an area of rotating thunderstorms with winds of 62 km/hr (38 mph) or less.

- A tropical depression becomes a tropical storm if its winds reach 63 km/hr (39 mph).

- A tropical storm becomes a hurricane if its winds reach 119 km/hr (74 mph).

Tornado

An F1 tornado in central Oklahoma

A tornado is a tube of violently spinning air that touches the ground. Wind inside the tornado spins fast, but the actual 'circle' of wind around them is huge. This makes tornadoes very dangerous.

Tornadoes are especially dangerous to people in cars or mobile homes and about 60 people are killed by tornadoes every year.

Tornadoes are devastating as they can cause significant injury and death. Tornadoes can tear through houses and often leave people homeless. Tornadoes can be caused by winds that have been going opposite directions with moist air meet as well as hurricanes. Nearly 3 Quarters of the worlds Tornadoes takes place in the US.

Tornadoes mostly happen during strong thunderstorms called super cell storms. They cause a lot of damage to anything in their path. Tornadoes are ranked on the Enhanced Fujita scale, from EF0 to EF5. EF0 for tornados that caused the least damage and EF5 for the ones that caused the most.

Tornadoes can happen in nearly any part of the world. In the United States, a tornado has happened in all states. The middle part of the United States is nicknamed 'Tornado Alley' for the number of tornadoes there. A tornado can have wind speeds of over 300 miles per hour (480 km/h). Most tornadoes have wind speeds less than 110 miles per hour (180 km/h), are about 250 feet (80 m) across and travel a few miles before disappearing. Other tornado-like phenomena that exist in nature include the gustnado, dust devil, fire whirls, and steam devil; downbursts are frequently confused with tornadoes, though their action is not similar.

Characteristics

Condensation Funnel

A tornado does not necessarily need to be visible; however, the extremely low pressure caused by the high wind speeds and rapid rotation usually causes water vapor in the air to condense into a visible *condensation funnel*. The tornado is the vortex of wind, not the condensation cloud.

Tornado Family

A single storm may produce multiple tornadoes and mesocyclones. Tornadoes produced from the same storm are referred to as a *tornado family*. Sometimes multiple tornadoes from distinct mesocyclones occur at the same time.

Tornado Outbreak

Occasionally, several tornadoes are spawned from the same very large storm. If there is no break in their activity, this is considered a *tornado outbreak*, although there are various definitions. A period of several successive days with tornado outbreaks in the same general area (spawned by multiple weather systems) is a *tornado outbreak sequence*, occasionally called an extended tornado outbreak.

Severe Tornado Outbreaks

Sometimes, tornadoes happen in groups. 148 tornadoes struck on the same day in April 1974. Many towns in the midwestern United States and Canada were destroyed. More than 300 people died. They were hit by flying wrecks, buried under houses, and thrown by powerful winds. That day, students in Xenia, Ohio were practicing for a play on the auditorium stage. One girl looked out the window and saw the tornado. The students ran into the hall, covering their heads. A few seconds later, all the school buses flew right onto the stage.

A man in another town hid under the couch in his living room. He held onto one couch leg. The tornado struck his house, and winds blew around him. When the tornado left, he was outside. There was no house. The couch had disappeared, and he was only holding onto one couch leg.

Tornado Watches/Warnings/Emergencies

A "tornado watch" is given when the weather conditions look like a tornado could form. A 'PDS (Particularly Dangerous Situation)' watch is given when a likely tornado outbreak is to start, many strong tornadoes will form in the area, or an ongoing tornado outbreak is in the works in the area. A "tornado warning" is given if somebody has actually seen a tornado or if a tornado 'signature' (usually the storm has a 'hook' or 'U' echo) has shown up on radar. Tornado emergencies are issued in Special Weather Statements or Tornado Warnings saying that a powerful tornado is about to hit an area with a lot of people in it (especially cities in Tornado Alley), a tornado has been spotted, and the tornado is expected to cause deaths.

Safety Tips

To keep safe in a tornado, here are some tips you can follow:

- Go to the lowest floor of the building. Stay close to the center of the building and away from windows, for example, a bathroom with no windows and get into the bathtub.

- Find a piece of strong furniture or a mattress to go under or hide in a closet and wait until it is over.

- If you are in a school, do *not* go to the gymnasium or any other place that has a high ceiling. Squat near the wall, placing your hands on the back of your head.

- If you cannot find shelter, find the lowest, most protected ground and cover your head with your hands.

- Do not drive in tornadoes. If you are in your car, position your head above the steering wheel, and cover yourself up.

- Do not seek shelter underneath an underpass or a bridge, as winds can send debris in your path.

Flood

Flood is a natural event or occurrence where a piece of land (or area) that is usually dry land, suddenly gets submerged under water. Some floods can occur suddenly and recede quickly. Others take days or even months to build and discharge.

When floods happen in an area that people live, the water carries along objects like houses, bridges, cars, furniture and even people. It can wipe away farms, trees and many more heavy items.

Floods occur at irregular intervals and vary in size, duration and the affected area.

It is important to note that water naturally flows from high areas to low lying areas. This means low-lying areas may flood quickly before it begins to get to higher ground.

Causes of Flood

Here are a few events that can cause flooding:

Rains

Each time there are more rains than the drainage system can take, there can be floods. Sometimes, there is heavy rain for a very short period that results in floods. In other times, there may be light rain for many days and weeks and can also result in floods.

River Overflow

Rivers can overflow their banks to cause flooding. This happens when there is more water up-stream than usual and as it flows downstream to the adjacent low-lying areas (also called a flood-plain), there is a burst and water gets into the land.

Strong Winds in Coastal Areas

Sea water can be carried by massive winds and hurricanes onto dry coastal lands and cause flooding. Sometimes this is made worse if the winds carry rains themselves. Sometimes water from the sea resulting from a tsunami can flow inland to cause damage.

Dams are man-made blocks mounted to hold water flowing down from a highland. The power in the water is used to turn propellers to generate electricity. Sometimes, too much water held up in the dam can cause it to break and overflow the area. Excess water can also be intentionally released from the dam to prevent it from breaking and that can also cause floods.

Ice and Snow-melts

In many cold regions, heavy snow over the winter usually stays un-melted for sometime. There are also mountains that have ice on top of them. Sometimes the ice suddenly melts when the temperature rises, resulting in massive movement of water into places that are usually dry. This is usually called a snowmelt flood.

Types of Floods

Some would like to see the causes of floods as types of floods, but on this page we shall look at three major flood types: Flash floods, Rapid on-set floods and slow on-set floods.

Flash Floods

This kind occurs within a very short time (2-6 hours, and sometimes within minutes) and is usually as a result of heavy rain, dam break or snow melt. Sometimes, intense rainfall from slow moving thunderstorms can cause it. Flash floods are the most destructive and can be fatal, as people are usually taken by surprise. There is usually no warning, no preparation and the impact can be very swift and devastating.

Rapid On-set Floods

Similar to flash floods, this type takes slightly longer to develop and the flood can last for a day or two only. It is also very destructive, but does not usually surprise people like Flash floods. With rapid on-set floods, people can quickly put a few things right and escape before it gets very bad.

Slow On-set Floods

This kind is usually as a result of water bodies over flooding their banks. They tend to develop slowly and can last for days and weeks. They usually spread over many kilometers and occur more in flood plains (fields prone to floods in low-lying areas). The effect of this kind of floods on people is more likely to be due to disease, malnutrition or snakebites.

Areas at Risk

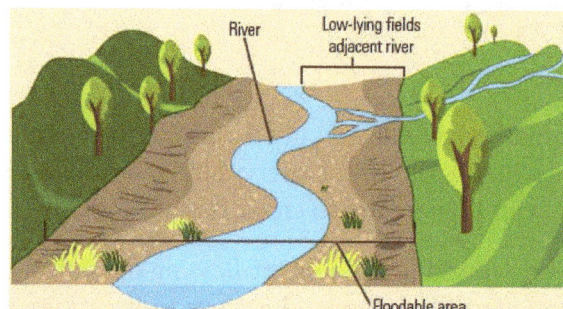

Generally, the natural behavior of water (and flowing water) is that it moves from higher ground to lower ground. This means if there is a higher ground adjacent a lower ground, the lower ground is a lot more likely to experience floods.

Additionally, anywhere that rains fall, floods can develop. This is so because anytime there are more rains bringing more water than it can be drained or absorbed by the soil, there is a flood potential.

In many cities, there are buildings springing up in many places where they have not been authorized. Some of these building are placed in waterways. Other places also have very bad and chocked drainage systems. The danger is that, with the rains, water will find its own level if it cannot find its way. The result is flooding and your home could be under water.

Any plain low-lying area adjacent a river, lagoon or lake is also more likely to have floods anytime the water level rises. This includes coastal areas and shorelines, as seawater can easily be swept inland by strong winds, tides and tsunamis.

Effects of Flooding

Floods can have devastating consequences and can have effects on the economy, environment and people.

Economic

During floods (especially flash floods), roads, bridges, farms, houses and automobiles are destroyed. People become homeless. Additionally, the government deploys firemen, police and other emergency apparatuses to help the affected. All these come at a heavy cost to people and the government. It usually takes years for affected communities to be re-built and business to come back to normalcy.

Environment

The environment also suffers when floods happen. Chemicals and other hazardous substances end up in the water and eventually contaminate the water bodies that floods end up in. In 2011, a huge tsunami hit Japan, and sea water flooded a part of the coastline. The flooding caused massive leakage in nuclear plants and has since caused high radiation in that area. Authorities in Japan fear that Fukushima radiation levels are 18 times higher than even thought.

Additionally, flooding causes kills animals, and others insects are introduced to affected areas, distorting the natural balance of the ecosystem.

People and Animals

Many people and animals have died in flash floods. Many more are injured and others made homeless. Water supply and electricity are disrupted and people struggle and suffer as a result. In addition to this, flooding brings a lot of diseases and infections including military fever, pneumonic plague, dermatopathia and dysentery. Sometimes insects and snakes make their ways to the area and cause a lot of havoc.

Methods of Flood Prevention

Humans cannot stop the rains from falling or stop flowing surface water from bursting its banks. These are natural events, but we can do something to prevent them from having great impact. Here are a few:

Sea/Coastal Defense Walls

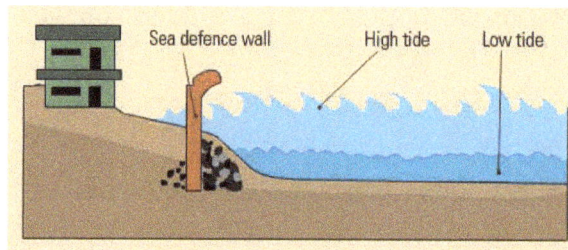

Sea walls and tide gates have been built in some places to prevent tidal waves from pushing the waters up ashore. In some areas too, sand bags are made and placed in strategic areas to retain floodwaters.

Retaining Walls

In some places, retaining walls levees, lakes, dams, reservoirs or retention ponds have been constructed to hold extra water during times of flooding.

Town Planning

It is important that builders acquire permission before buildings are erected. This will ensure that waterways are not blocked. Also, drainage systems must be covered and kept free from objects that chock them. This way, water can quickly run through if it rains and minimize any chance of town flooding. Drainage systems should also be covered to prevent litter from getting into them.

Vegetation

Trees, shrubs and grass help protect the land from erosion by moving water. People in low-lying areas must be encouraged to use a lot of vegetation to help break the power of moving flood water and also help reduce erosion.

Education

In many developing countries, drainage systems are chocked with litter and people have little knowledge of the effects that can have during a rain. When it rains, waterways and culverts are blocked by massive chunks of litter and debris, and water finds its way into the streets and into people's homes. Education is therefore very important, to inform and caution people about the dangers of floods, what causes floods, and what can be done to minimize its impact.

Detention Basin

These are small reservoirs built and connected to waterways. They provide a temporary storage for floodwaters. This means in an event of flooding, water is drained into the basin first, giving people more time to evacuate. It can also reduce the magnitude of downstream flooding.

Droughts

A drought is an extended period of months or years when a region notes a deficiency in its water supply. Generally, this occurs when a region receives consistently below average precipitation. It can have a substantial impact on the ecosystem and agriculture of the affected region. Although droughts can persist for several years, even a short, intense drought can cause significant damage and harm the local economy.

Implications

Drought is a normal, recurring feature of the climate in most parts of the world. Having adequate drought mitigation strategies in place can greatly reduce the impact. Recurring or long-term drought can bring about desertification. Recurring droughts in the Horn of Africa have created grave ecological catastrophes, prompting massive food shortages, still recurring. To the northwest of the Horn, the Darfur conflict in neighboring Sudan, also affecting Chad, was fueled by decades of drought; combination of drought, desertification and overpopulation are among the causes of the Darfur conflict, because the Arab Baggara nomads searching for water have to take their livestock further south, to land mainly occupied by non-Arab farming peoples.

According to a UN climate report, the Himalayan glaciers that are the sources of Asia's biggest rivers—Ganges, Indus, Brahmaputra, Yangtze, Mekong, Salween and Yellow—could disappear by 2035 as temperatures rise. Approximately 2.4 billion people live in the drainage basin of the Himalayan Rivers. India, China, Pakistan, Bangladesh, Nepal and Myanmar could experience floods followed by droughts in coming decades. Drought in India affecting the Ganges is of particular concern, as it provides drinking water and agricultural irrigation for more than 500 million people. Paradoxically, some proposed short-term solutions to global warming also carry with them increased chances of drought.

In 2005, parts of the Amazon basin experienced the worst drought in 100 years. In July 2006 Woods Hole Research Center results showing that the forest in its present form could survive only three years of drought. Scientists at the Brazilian National Institute of Amazonian Research argue that this drought response, coupled with the effects of deforestation on regional climate, are pushing the rainforest towards a "tipping point" where it would irreversibly start to die. It concludes that the rainforest is on the brink of being turned into savanna or desert, with catastrophic consequences for the world's climate. According to the WWF, the combination of climate change and deforestation increases the drying effect of dead trees that fuels forests fires.

Causes

Generally, rainfall is related to the amount of water vapor in the atmosphere, combined with the upward forcing of the air mass containing that water vapor. If either of these is reduced, the result is drought.

Factors include:

- Above-average prevalence of high pressure systems.

- Winds carrying continental, rather than oceanic air masses (ie. reduced water content).

- El Nino (and other oceanic temperature cycles).

- Deforestation.

- Stages of drought.

As a drought persists, the conditions surrounding it gradually worsen and its impact on the local population gradually increases. Droughts go through three stages before their ultimate cessation:

1. Meteorological drought is brought about when there is a prolonged period with less than average precipitation. Meteorological drought usually precedes the other kinds of drought.

2. Agricultural droughts are droughts that affect crop production or the ecology of the range. This condition can also arise independently from any change in precipitation levels when soil conditions and erosion triggered by poorly planned agricultural endeavors cause a shortfall in water available to the crops. However, in a traditional drought, it is caused by an extended period of below average precipitation.

3. Hydrological drought is brought about when the water reserves available in sources such as aquifers, lakes and reservoirs falls below the statistical average. Like an agricultural drought, this can be triggered by more than just a loss of rainfall. For instance, Kazakhstan was recently awarded a large amount of money by the World Bank to restore water that had been diverted to other nations from the Aral Sea under Soviet rule. Similar circumstances also place their largest lake, Balkhash, at risk of completely drying out.

Consequences

Periods of drought can have significant environmental, economic and social consequences. The most common consequences include:

- Death of livestock.

- Reduced crop yields.

- Wildfires, such as Australian bushfires, are more common during times of drought.

- Shortages of water for industrial users.

- Desertification.

- Dust storms, when drought hits an area suffering from desertification and erosion.

- Malnutrition, dehydration and related diseases.

- Famine due to lack of water for irrigation.

- Social unrest.

- Mass migration, resulting in internal displacement and international refugees.

- War over natural resources, including water and food.

- Reduced electricity production due to insufficient available coolant.

- Snakes have been known to emerge and snakebites become more common.

The effect varies according to vulnerability. For example, subsistence farmers are more likely to migrate during drought because they do not have alternative food sources. Areas with populations that depend on subsistence farming as a major food source are more vulnerable to drought-triggered famine. Drought is rarely if ever the sole cause of famine; socio-political factors such as extreme widespread poverty play a major role. Drought can also reduce water quality, because lower water flows reduce dilution of pollutants and increase contamination of remaining water sources.

Drought Mitigation Strategies

- Desalination of sea water for irrigation or consumption.

- Drought monitoring: Continuous observation of rainfall levels and comparisons with current usage levels can help prevent man-made drought. For instance, analysis of water usage in Yemen has revealed that their water table (underground water level) is put at grave risk by over-use to fertilize their khat crop. Careful monitoring of moisture levels can also help predict increased risk for wildfires, using such metrics as the Keetch-Byram Drought Index or Palmer Drought Index.

- Land use: Carefully planned crop rotation can help to minimize erosion and allow farmers to plant less water-dependent crops in drier years.

- Rainwater harvesting: Collection and storage of rainwater from roofs or other suitable catchments.

- Recycled water - Former wastewater (sewage) that has been treated and purified for reuse.

- Transvasement: Building canals or redirecting rivers as massive attempts at irrigation in drought-prone areas.

- Water restrictions: Water use may be regulated (particularly outdoors). This may involve regulating the use of sprinklers, hoses or buckets on outdoor plants, the washing of motor vehicles or other outdoor hard surfaces (including roofs and paths), topping up of swimming pools, and also the fitting of water conservation devices inside the home (including shower heads, taps and dual flush toilets).

- Cloud seeding: an artificial technique to induce rainfall.

References

- Dylan J. Livengood (2004). "Tornado Outbreak Day Sequences: Historic Events and Climatology (1875–2003)" (PDF). Retrieved 2007-03-20

- Causes-effects-and-types-of-landslides, natural-disaster: eartheclipse.com, Retrieved 12 March 2018

- Types-of-volcanic-eruptions, volcanoes: geology.com, Retrieved 09 April 2018

- Oard, Michael (1997). The Weather Book. P.O. Box 126, Green Forest, AR 72638: Master Books. ISBN 0-89051-211-6

- Why-do-volcanoes-erupt, geoscience: cosmosmagazine.com, Retrieved 11 May 2018

- What-is-a-flood, natural-disasters: eschooltoday.com, Retrieved 18 April 2018

6

Hazardous Waste

Hazardous wastes are the wastes that pose a potential threat to public health and safety. It can also cause harm to the environment. The topics elaborated in this chapter on household hazardous waste, radioactive waste and waste treatment technology, have been carefully written to provide a comprehensive understanding of hazardous wastes.

Hazardous waste is legally defined as solid waste with the potential to harm humans or the environment. This kind of waste has to be taken care of very carefully for safety reasons, and there are special regulations related to handling and disposing it. Most hazardous wastes are poisonous in some way, but some are classified as hazardous because they are flammable or explosive. A lot of hazardous waste comes from industrial processes, and increased regulation has generally led to a reduction in the amount produced.

The protection agencies define four primary kinds of hazardous waste. The first type is called "listed wastes," and that basically means that they come from industrial or scientific processes, and the agencies have protocols to deal with them. "Universal wastes" are found in everyday items like batteries. "Characteristic wastes" are similar to listed wastes, but they aren't as well-documented, and "mixed wastes" are generally radioactive materials combined with other waste components.

The vast majority of hazardous materials come from businesses, and some aren't actually big industrial companies. For example, auto repair shops produce a lot of hazardous waste, as do hospitals. Generally speaking, companies that produce hazardous waste products are required by law to dispose of them in an appropriate manner and protect the public from exposure. Businesses that violate these rules can be subject to legal repercussions, including private lawsuits and government prosecutions.

In the early 1990s, laws were passed that made it illegal to use regular landfills for disposal of hazardous substances. The only exception was in cases where the waste had been chemically treated to make it less harmful. There has generally been a fair amount of industrial resistance to these kinds of laws, because they can raise manufacturing costs, but that has lessened over time as more efficient disposal methods have come to the market. Some companies have found ways to recycle hazardous waste products and turn them into useful substances, while others destroy the waste with incineration procedures and similar methods.

The average person can produce significant hazardous materials in her own home. For example, pesticides used in a person's garden can be classified as hazardous waste, and certain cleaning products are potentially very dangerous. One of the main sources of household hazardous waste comes from the automobile and various items used to maintain it. Gasoline, oil, antifreeze, and battery acid are all byproducts of automotive maintenance, and they can pose a significant risk.

Household Hazardous Waste

Household hazardous waste includes cleaners, stains, varnishes, batteries, automotive fluids, pesticides, herbicides, certain paints, and many other items found in basements, under kitchen sinks and dark garage corners.

Identification of Household Hazardous Waste

Household hazardous wastes fall into one of four categories, which are noted on the container.

- Flammable: Ignites easily and burns rapidly
- Corrosive: May cause deterioration of body tissues or erosion of material at the site of contact.
- Explosive: Contents may explode if incinerated or stored above 82 °F.
- Toxic/Poison: Harmful or deadly upon contact, ingestion or inhalation.

Dangers of Improper Disposal

Household hazardous wastes are sometimes disposed of improperly by individuals pouring wastes down the drain, on the ground, into storm sewers, or putting them out with the trash. Improperly discarded household hazardous wastes have the potential to cause physical injury to sanitation workers, contaminate septic tanks or waste water treatment systems and present hazards to children and pets.

Proper Disposal of Household Hazardous Waste

If you use products with hazardous components, purchase and use only the amount needed. Leftover materials can be shared with neighbors or donated to a charity, business, or government agency, or given to a household hazardous waste collection program.

If you are planning to move, don't leave your household hazardous waste for future residents. Remember to plan ahead and properly dispose of your hazardous waste.

Radioactive Waste

Radioactive waste is nuclear fuel that is produced after being used inside of a nuclear reactor. Although it looks the same as it did before it went inside of the nuclear producer it has changed compounds and is nothing like the same. What is left is considered radioactive material and is very dangerous to anyone. This is very dangerous and remains this way for not just a few years but for thousands of years. It must be handled in the right manner so not to cause a ton of devastation in the world. It could take just seconds to die from exposure to radioactive materials. In short, radioactive waste is a kind of waste in gas, liquid or solid form that contains radioactive nuclear substance.

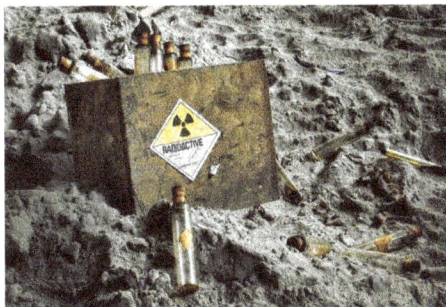

There are many industries like mining, defense, medicine, scientific research, nuclear power generation which produce by-products that include radioactive waste. The radioactive waste can remain radioactive for few months, years or even hundreds of years and the level of radioactivity can vary. The radioactive waste is extremely toxic as it can remain radioactive for so long and can cause acute radiation sickness when it first comes out of the reactor, if you stood within a few meters of it while it was unshielded.

It is very dangerous, although some types of radioactive waste are considered to be more harmful than other types. Since it is so hazardous and toxic, finding suitable disposal sites for radioactive waste remains a tedious task. Therefore, safe disposal site is required to ensure safety of humans, animals and this environment from toxic gases.

Types of Radioactive Waste

High-level Waste

You will find that there are two types of nuclear reactors. These types are the pressurized and boiler water reactors. High-level nuclear waste, simply put, is spent fuel that is still present after it has been used inside of nuclear reactors. This radioactive waste has to cool off for several years and is considered to be very dangerous. The cooling process of this waste usually takes place inside of deep pools of water that are several hundred feet deep. These pools can be located on-site of off-site of the nuclear facility although the off-site facilities are limited and must be approved by the EPA.

This type of waste is hazardous to people for many reasons, but especially because it remains radioactive. High level waste accounts for 95% of the total radioactivity produced in the nuclear reactor. This type of nuclear waste is very dangerous. It must consistently go through a process to keep it cool and the radioactive material under control. High level waste can have short and long lived components depending upon the time it will take for the radioactivity to decrease to levels that is not considered harmful for humans and surrounding environment.

Intermediate-level Waste

Intermediate-level waste contains high amount of radioactivity than low-level and less than high-level. This type of waste typically requires shielding during handling and interim storage. This type of waste typically includes refurbishment waste, ion-exchange resins, chemical sludges and metal fuel cladding. The intermediate level waste contains 4% of the entire radioactivity. Intermediate-level waste that requires long term management is transferred to an authorized waste management operator.

Low-level Waste

Most of the radioactive waste that is around today is considered to be low level. In fact, about 90% of all nuclear waste is low level. Nuclear reactors, hospitals, dental offices, and similar types of facilities often use low-level nuclear waste materials on a daily basis and it is needed in order to provide the services that are offered within these facilities. Low-level nuclear waste is not dangerous, and any of it can be disposed of inside of a landfill. This is the reason why it does not require shielding during handling and transport.

Even so there is a strict criterion in which it must be handled and disposed of. Without the proper disposal what is not dangerous has the possibilities of becoming that way. This is not a chance that you should be willing to take when it is so very easy to protect yourself. The low level waste contains just 1% of the radioactivity of all radioactive waste.

Mining and Milling

Tailings and waste rock are generates by mining and milling of uranium ore. The tailings material is covered with water and has the consistency of fine sand, when dried. It is produced by grinding the ore and the chemical concentration of uranium. After few months, the tailings material contains 75% of the radioactivity of the original ore.

Clean and mineralized waste rock is produced during mining activities which must be excavated to access to access uranium ore body. It has little or no concentration of uranium. While clean waste rock can be used for construction purposes mineralized waste rock could generate acid when left on the surfaced indefinitely that could affect surrounding environment.

Transuranic Waste

Transuranic waste, or TRU waste contains more than 3700 be per gram of elements. It is much heavier than uranium. This type of waste is produced through nuclear waste reprocessing procedures in most cases. This is one of the least worried about types of radioactive waste that is out there but it is worth mentioning since it is a part of nuclear waste.

Other Classifications

Non-commercial activities can bring nuclear waste to the forefront. There are several activities, including by-product materials and uranium mining, to name a few. The clean-up standards and procedures for these activities vary and are set forth by the Environmental Protection Agency.

Waste Treatment Technology

Technologies used for managing and treating waste are known as waste treatment technologies. They are usually used for the storage, disposal and recycling of different types of waste. Waste treatment technologies are used to treat different kinds of waste, like solid waste, liquid waste, hazardous waste etc.

Incineration

A waste treatment technology, which includes the combustion of waste for recovering energy, is called as "incineration". Incineration coupled with high temperature waste treatments are recognized as thermal treatments. During the process of incineration, the waste material that is treated is converted in to IBM, gases, particles and heat. These products are later used for generation of electricity. The gases, flue gases are first treated for eradication of pollutants before going in to atmosphere.

Among waste-to-energy technologies, incineration stands taller. Other technologies are gasification, PDG, anaerobic digestion and pyrolysis. Sometimes Incineration is conducted without the reason for recovering energy.

In past, incineration was conducted without separating materials thus causing harm to environment. This un-separated waste was not free from bulky and recyclable materials, even. This resulted in risk for plant workers health and environment. Most of such plants and incinerations never generate electricity.

Incineration reduces the mass of the waste from 95 to 96 percent. This reduction depends upon the recovery degree and composition of materials. This means that incineration however, does not replace the need for landfilling but it reduced the amount to be thrown in it.

Incineration comes with a number of benefits in specific areas like medical wastes and other life risking waste. In this process, toxins are destroyed when waste is treated with high temperature.

Incineration or thermal treatment of waste is much popular in countries like Japan where there is scarcity of land. The energy generated by incineration is highly demanded in countries like Denmark and Sweden. In year, 2005 it was estimated that 4.8 percent of the electricity as is consumed by Danish nation was produced by incineration and the amount of heat was some 13.7 percent out of total. Other than Denmark and Sweden many European countries are recovering heat and electricity from waste.

Technology

Incinerators and their Types

Incinerator can be understood more precisely as a furnace where waste is burnt. Modern incinerators are equipped with pollution improvement systems, which play their part in cleaning up the Flue gas and such toxicants. Following are the types of plants for burning waste:

Moving Grate

The incineration plant used for treating MSW is moving grate. This grate is capable for hauling waste from combustion chamber to give way for complete and effective combustion. A single such plant is capable for taking in thirty-five metric ton of waste every hour for treatment. Moving grates are more precisely known as incinerators of municipal solid waste.

This waste is poured in the grate with a help of crane from and opening or throat. From here, the waste has to move towards the ash pit. Waste is further treated and water locks wash out ash from it. Air is then flown through the waste and this blown air works for cooling down the grate. Some of grates are cooled with help of water.

Air is blown through the boiler for another time but this time comparatively faster than before. This air helps in complete burning of the flue gases with the introduction of turmoil leading to better mixing and excess of oxygen. In some grates, the combustion air at fast speed is blown in separate chamber.

European Waste incineration Directive is of the view that an incineration plant must be designed so that operating worker must know that flue gases are reaching the temperature of eight fifty degrees centigrade with in two seconds. This would ensure complete and required breakdown of toxins of organic nature. In order to achieve this every time backup auxiliary burners must be installed.

Fixed Grate

This was the fixed and much older version for grate. This kind generally is lined with the brick while lower or ash pit is made up of metal. This grate generally has an opening at the top and for loading purpose; a side of the grate is left open. A number of fixed grate were first formed in houses, which today are replaced by waste compactors.

Rotary-kiln

Industries and municipalities generally use this sort of incinerator. This incinerator consists of two chambers i.e. primary and secondary chamber.

Fluidized Bed

In this sort of incineration, air is blown at high speed over a sand bed. The air gets going through the bed when a point come where sand granules separates and let air pass through them and here comes the part of mixing and churning. Therefore, a fluidized bed comes in to being and fuel and waste are then can be introduced.

The sand along with the pretreated fuel or waste is kept suspended and is pumped through the air currents. The bed is thus mixed violently and is uptight while small inert particles are kept suspended in air in form of fluid like form. This let the volume of the waste, sand and fuel to be circulated throughout the furnace, completely.

Specialized Incineration

When it comes to the furniture factory for incineration of the waste, they need to take special precautions, as they have to handle inflammable material. For this purpose, they have incinerators, which are installed with burn back prevention systems and are very much necessary for the dust suspensions when they are more able to catch up the fire.

Use of Heat

The heat that is produced by an incinerator can be used for generating steam, which is used for driving a turbine in order to produce electricity. The typical amount as is produced by Municipal waste per ton is 2/3 MWh for electricity and two MWh for heating.

Pollution

Incineration is conducted with a number of outputs, which include ash and flue gas emission. Before the flue gas cleaning systems were introduced, the flue gas has to move to atmosphere thus leading to pollution.

Emission of Gases

Furans and Dioxins

The biggest most concern, which has caught thoughts of environmentalists about MSW's incineration, is production of a huge amount of furans and dioxins. These are considered staidly injurious to health. Modern generators are equipped with special equipment to clean emission of gases from these injurious components. There was a time when no governmental regulation were there to bound incineration and save environment and atmosphere from this hazardous emission of gases but today there are strict and rigid rules and regulations to follow and conduct incineration.

Carbon Dioxide

Incineration while being conducted produces a vast amount of Carbon dioxide. Carbon dioxide plays a due role in global warming, as this is the greenhouse gas. It has been observed that almost everything which has carbon in its composition is when processed by incineration evolves out as carbon dioxide.

Extra Emissions

Some other emissions of gases by waste processing are sulfur dioxide, hydrochloric acid, fine particles and heavy metals.

Cleaning out Flue Gas

A number of processes are involved for the cleaning up of flue gas.

The mixture of flue gas is collected by means of Particle filtration and this filtration is conducted using electrostatic precipitators and baghouse filters. Baghouse are very effective for fine particles. The next step of the processing and cleaning of flue gas is processing of scrubbers, which are critical for the removal of hydrochloric acid, nitric acid, mercury, hydrofluoric acid, lead and residuary heavy metals. With the reaction of lime, sulfur is converted in to gypsum. The wastewater, which comes out of scrubbers, is then passed through wastewater treatment plant.

Desulphurization is a process that is used to remove sulfur dioxide with the limestone slurry injection directly in to flue gas. Nitric component or gases are reduced with catalytic reduction with help of ammonia application. Heavy metals are removed with the help of active carbon injection. Particles are the collected at filters.

Solidify Outputs

Flue ash and Bottom ash is produced with the processing of waste materials and settle at the bottom of the incineration plant. The ash, which is produced, is four to five percent of total weight of the waste processed while the flue ash makes up some ten to twenty percent of total weight of waste material. The heavy metals, which are contained in the flue or bottom ash, are lead, cadmium, zinc and copper. A small amount of furans and dioxins are also produced. It is to mention here that bottom ash seldom have heavy metals in it. Flue ash is hazardous while bottom ash is not that dangerous or injurious to health.

Other Issues Related to Pollution

Older models of incinerators have inconvenience that this produce odor pollution. However, in modern plants are saved from producing dust and odor pollution. They are designed to store waste in enclosed containers along with a negative pressure to keep from odor and dirt dispersal.

Another issue that is affecting community is increased load of traffic due to WCV for hauling waste materials. This is the issue, which has forced incinerators to move in to industrial areas.

Pyrolysis

Pyrolysis is a thermochemical treatment, which can be applied to any organic (carbon-based) product. In this treatment, material is exposed to high temperature, and in the absence of oxygen goes through chemical and physical separation into different molecules.

This allows receiving products with a different, often more superior character that original residue. Thanks to this feature, pyrolysis becomes increasingly important process for today industry – as it allows bringing far greater value to common materials and waste.

In contrary to combustion and gasifications processes, which involve entire or partial oxidation of material, pyrolysis bases on heating in the absence of air. This makes it mostly endothermic process that ensures high energy content in the products received.

Pyrolysis products always produce solid (charcoal, biochar), liquid and non-condensable gases $(H_2, CH_4, CnHm, CO, CO_2$ and $N)$. As the liquid phase is extracted from pyrolysis gas only during its cooling down, in some applications, these two streams can be used together when providing hot syngas directly to the burner or oxidation.

During the pyrolysis, a particle of material is heated up from the ambient to defined temperature. The material remains inside the pyrolysis unit and is transported by screw conveyor at defined speed, until the completion of the process. Chosen temperature of pyrolysis defines the composition and yields of products (pyrolysis oil, syngas and char).

Factor of Influences Pyrolysis Process Results

Treated material composition: each of the major constituents of biomass and waste feature different temperatures of thermal decomposition, which means they contribute to the results of process in different way. Due to high diversity of material compositions it is always recommended to perform a pilot tests to forecast the pyrolysis process performance in most accurate way.

Temperature of process: has a major influence to the treatment results. Higher temperatures of pyrolysis provide greater quantity of non-condensable gases (syngas, synthetic gas), while lower temperatures favours the production of high quality solid product (charcoal, bio-coal, torrefied fuels). Temperature is a factor fully controllable in Biogreen® process thanks to electrically heated screw conveyor that allows precise setup of treatment conditions.

Residence time of material in the pyrolysis chamber: influences the degree of thermal conversion of received solid product as well as the residence time of the vapor, which influences the composition of vapors (condensable / non-condensable phase). Residence time can be precisely controlled in Biogreen® process by changing the rotation speed of screw conveyor transporting material along the reactor.

Particle size and physical structure: influences the speed in which material is subjected to pyrolysis. In general, lower particle size materials are quicker affected by the thermal decomposition, which can result in greater quantities of pyrolysis oil than in case of larger particle size.

Solidification/Stabilization with Cement

Solidification/stabilization techniques are akin to locking the contaminants in the soil. It is a process that physically encapsulates the contaminant. This technique can be used alone or combined with other treatment and disposal methods.

The most common form of S/S is a cement process. It simply involves the addition of cement or a cement-based mixture, which thereby limits the solubility or mobility of the waste constituents. These techniques are accomplished either in situ, by injecting a cement based agent into the contaminated materials or ex situ, by excavating the materials, machine-mixing them with a cement-based agent, and depositing the solidified mass in a designated area. The goal of the S/S process is to limit the spread, via leaching, of contaminated material. The end product resulting from the solidification process is a monolithic block of waste with high structural integrity. Types of solidifying/stabilizing agents include the following: Portland; gypsum; modified sulfur cement, consisting of elemental sulfur and hydrocarbon polymers; and grout, consisting of cement and other dry materials, such as acceptable fly ash or blast furnace slag. Processes utilizing modified sulfur cement are typically performed ex situ.

The Department of Energy (DOE) is developing another system called the Polyethylene Encapsulation of Radionuclides and Heavy Metals (PERM) process. This process encapsulates contaminants in polyethylene. It is used for radionuclides (e.g., cesium, strontium, and cobalt) and toxic metals (e.g., chromium, lead, and cadmium). Most S/S products are designed to be left in place, although it is possible that the solid materials could be moved to other locations.

Limitations and Concerns

The depth of contaminants may limit these processes.

Future use of the site and environmental conditions may erode the materials used to stabilize contaminants, thus affecting their capacity to immobilize contaminants. Solidified material may also restrict future use of the site.

Very little data exist to support S/S products durability over their indefinite disposal life. Long term monitoring is necessary to ensure that contaminants have not been re-mobilized.

Certain waste streams are incompatible with variations of this process, and each application must be carefully tested for long-term compatibility before it is used.

Organic wastes are generally not immobilized, and unless very high temperatures are used to destroy them, they will migrate. However, if a process were designed to destroy organic compounds through heating, the creation of products of incomplete combustion such as dioxin and furan would pose a concern.

Inorganic salts affect the set rate either through acceleration or retardation. Users need to know precisely how different salts individually and collectively affect basic Portland cement stabilization so the proper additive can be used in the dry binder mix.

When radioactive contamination is present, other types of hazardous waste (e.g., organic chemicals) may interfere with solidification. Treatability studies are needed to demonstrate that the S/S process works.

Given the long period of time that radioactive waste will be a hazard, the S/S facility must be particularly careful about the degradation of construction materials. Current research has focused on developing new types of materials to improve liner integrity and to reduce possible radionuclides migration.

For radioactive waste, there is concern about the likelihood of liner deterioration, liner penetration, and leaching over the long-term, as well as risks associated with the possible excavation, handling, and transportation of radioactive waste.

In situ S/S may not be suitable for some sites because gamma radiation might not be reduced sufficiently.

With in situ S/S, consideration must be given to any debris such as barrels, metal scrap, and wood pieces that may interfere with the solidification process.

Soil characteristics influence whether the technology will contain the waste effectively. These characteristics include void volume, which determines how much grout can be injected into the site; soil pore size, which determines the size of the cement particles that can be injected; and permeability of the surrounding area, which determines whether water will flow preferentially around the solidified mass.

Some cement processes result in significant increases in volume— up to double the original quantity.

In ex-situ applications of the cement S/S, volatile organic compounds (VOCs) are generally not immobilized.

In ex-situ applications, cracks extending through the stabilized mass have been observed, the cause of which is suspected to be the high temperature rise during curing.

Applicability

The target contaminant group for physical S/S is generally inorganics (including radionuclides) in the soil. While it may be effective for some organics, this technology may have limited effectiveness with semi volatile organic compounds (SVOCs) and pesticides. Encapsulation is also used for low-level radioactive mixed waste.

Technology Development Status

Physical stabilization techniques are well documented and commercialized. DOE has demonstrated the PERM process at the bench scale.

References

- What-is-hazardous-waste: wisegeek.com, Retrieved 12 March 2018

- Incineration: wrfound.org.uk, Retrieved 16 July 2018

- What-is-pyrolysis: biogreen-energy.com, Retrieved 20 April 2018

- Types-of-radioactive-waste: conserve-energy-future.com, Retrieved 11 June 2018

- What-is-considered-household-hazardous-waste, household-hazardous-waste: acwastewatcher.org, Retrieved 16 July 2018

7

Pollution

The introduction of any form of contaminant into the natural environment can cause harm to the environment and to life forms. This is termed as pollution. It can be of different forms. The topics covered in this chapter address the varied types of pollution such as air, water, noise and soil pollution and the different mitigation strategies for each.

Pollution is the introduction of harmful materials into the environment. These harmful materials are called pollutants. Pollutants can be natural, such as volcanic ash. They can also be created by human activity, such as trash or runoff produced by factories. Pollutants damage the quality of air, water, and land.

Pollution is a global problem. Although urban areas are usually more polluted than the countryside, pollution can spread to remote places where no people live. For example, pesticides and other chemicals have been found in the Antarctic ice sheet. In the middle of the northern Pacific Ocean, a huge collection of microscopic plastic particles forms what is known as the Great Pacific Garbage Patch.

Types & Causes of Pollution

Air Pollution is the most prominent and dangerous form of pollution. It occurs due to many reasons. Excessive burning of fuel which is a necessity of our daily lives for cooking, driving and other industrial activities; releases a huge amount of chemical substances in the air every day; these pollute the air.

Smoke from chimneys, factories, vehicles or burning of wood basically occurs due to coal burning; this releases sulphur dioxide into the air making it toxic. The effects of air pollution are evident too. Release of sulphur dioxide and hazardous gases into the air causes global warming and acid rain; which in turn have increased temperatures, erratic rains and droughts worldwide; making it tough for the animals to survive. We breathe in every polluted particle from the air; result is increase in asthma and cancer in the lungs.

Water Pollution has taken toll of all the surviving species of the earth. Almost 60% of the species live in water bodies. It occurs due to several factors; the industrial wastes dumped into the rivers and other water bodies cause an imbalance in the water leading to its severe contamination and death of aquatic species. If you suspect that nearby water sources have been contaminated by a corporation then it might be a good idea to hire an expert to see your options.

Also spraying insecticides, pesticides like DDT on plants pollutes the ground water system and oil spills in the oceans have caused irreparable damage to the water bodies. Eutrophication is another big source; it occurs due to daily activities like washing clothes, utensils near lakes, ponds or rivers; this forces detergents to go into water which blocks sunlight from penetrating, thus reducing oxygen and making it inhabitable.

Water pollution not only harms the aquatic beings but it also contaminates the entire food chain by severely affecting humans dependent on these. Water-borne diseases like cholera, diarrhea have also increased in all places.

Soil pollution occurs due to incorporation of unwanted chemicals in the soil due to human activities. Use of insecticides and pesticides absorbs the nitrogen compounds from the soil making it unfit for plants to derive nutrition from. Release of industrial waste, mining and deforestation also exploits the soil. Since plants can't grow properly, they can't hold the soil and this leads to soil erosion.

Noise pollution is caused when noise which is an unpleasant sound affects our ears and leads to psychological problems like stress, hypertension, hearing impairment, etc. It is caused by machines in industries, loud music, etc.

Radioactive pollution is highly dangerous when it occurs. It can occur due to nuclear plant malfunctions, improper nuclear waste disposal, accidents, etc. It causes cancer, infertility, blindness, and defects at the time of birth; can sterilize soil and affect air and water.

Thermal/heat pollution is due to the excess heat in the environment creating unwanted changes over long time periods; due to huge number of industrial plants, deforestation and air pollution. It increases the earth's temperature, causing drastic climatic changes and extinction of wildlife.

Light pollution occurs due to prominent excess illumination of an area. It is largely visible in big cities, on advertising boards and billboards, in sports or entertainment events at the night. In residential areas the lives of the inhabitants is greatly affected by this. It also affects the astronomical observations and activities by making the stars almost invisible.

Effects of Pollution

1. Environment degradation: Environment is the first casualty for increase in pollution weather in air or water. The increase in the amount of CO_2 in the atmosphere leads to smog which can restrict sunlight from reaching the earth. Thus, preventing plants in the process of photosynthesis. Gases like Sulfur dioxide and nitrogen oxide can cause acid rain. Water pollution in terms of Oil spill may lead to death of several wildlife species.

2. Human health: The decrease in quality of air leads to several respiratory problems including asthma or lung cancer. Chest pain, congestion, throat inflammation, cardiovascular disease, respiratory diseases are some of diseases that can be causes by air pollution. Water pollution occurs due to contamination of water and may pose skin related problems including skin irritations and rashes. Similarly, Noise pollution leads to hearing loss, stress and sleep disturbance.

3. Global warming: The emission of greenhouse gases particularly CO2 is leading to global warming. Every other day new industries are being set up, new vehicles come on roads and trees are cut to make way for new homes. All of them, in direct or indirect way lead to increase in CO2 in the environment. The increase in CO2 leads to melting of polar ice caps which increases the sea level and pose danger for the people living near coastal areas.

4. Ozone layer depletion: Ozone layer is the thin shield high up in the sky that stops ultra violet rays from reaching the earth. As a result of human activities, chemicals, such as chlorofluorocarbons (CFCs), were released int to the atmosphere which contributed to the depletion of ozone layer.

5. Infertile land: Due to constant use of insecticides and pesticides, the soil may become infertile. Plants may not be able to grow properly. Various forms of chemicals produced from industrial waste is released into the flowing water which also affects the quality of soil.

Pollution not only affect humans by destroying their respiratory, cardiovascular and neurological systems; it also affects the nature, plants, fruits, vegetables, rivers, ponds, forests, animals, etc, on which they are highly dependent for survival. It is crucial to control pollution as the nature, wildlife and human life are precious gifts to the mankind.

Pollutants

A pollutant is the major part or component of Pollution, the process by which natural environments are damaged by artificial or harmful objects, actions, or neglect. The pollutant itself might only be damaging in large numbers, or it could cause major toxic damage from only a small dose.

Environmental pollutants can be derived from a number of sources. Knowing what the different types of pollution are and where they come from can help you to understand the potential impact of these pollutants on your health and the health of the planet.

Soil Pollutants

Soil pollution is the pollution of the Earth's land surfaces, the most common types of soil pollutants are heavy metals such as cadmium, chromium, copper, zinc or mercury, pesticides or herbicides, organic chemicals, oils and tars, explosive or toxic gases, combustible or radioactive materials, biologically active compounds and asbestos. These types of pollutants can enter the soil through poor agricultural practices, industrial runoff, mining, landfill leakage, littering or the improper or illegal dumping of household or industrial waste materials.

Air Pollutants

Air pollution is the pollution of the Earth's atmosphere. There are six types of common air pollutants. They include ozone, particulate matter, carbon monoxide, nitrogen oxides, sulfur dioxide and lead. These and other air pollutants typically enter the atmosphere through industrial processes related to the generation of heat and power, incineration of solid wastes and transportation, emissions from vehicles are estimated to be responsible for approximately 60% of all air pollution alone and 80% of air pollution in cities.

Water Pollutants

Water pollution is the pollution of the Earth's oceans and other water sources, common types of water pollutants include mercury, nitrates, phosphorous, fecal coliform and bacterial pollution. These and other types of pollutants enter the water supply through industrial waste runoff, sewage treatment plants, feedlots, urban and agricultural runoff, septic systems and the illegal dumping of solid waste.

Noise Pollutants

Noise pollution is a form of air pollution related specifically to the types of sound present in the atmosphere. The Environmental Protection Agency defines a noise pollutant as any sound that interferes with normal activities or disrupts or diminishes one's quality of life. Noise pollutants can be present in the home, school, work or the community at large. Different types of noise pollutants may include sounds generated by aircraft, trains, boats, automobile traffic, construction, industrial manufacturing, vehicle alarms or even loud music.

Water Pollution

Water pollution is defined as the presence in groundwater of toxic chemicals and biological agents that exceed what is naturally found in the water and may pose a threat to human health and/or the environment. Additionally, water pollution may consist of chemicals introduced into the water bodies as a result of various human activities. Any amount of those chemicals pollutes the water, regardless of the harm they may pose to human health and the environment.

Causes

Many of the chlorinated solvents commonly used in industry (such as PCE, TCE, 1,1,1-TCA) are examples of such chemicals polluting our waters exclusively due to human activities. Another example is MTBE (Methyl-tert-butyl-ether), a gasoline oxygenate that is currently banned in the U.S. Regardless of their provenance, the chemicals or biological agents causing water pollution are generically referred to as water pollutants. The chemical and biological agents represent the main causes of water pollution and are generically referred to as water pollutants.

Any kind of water can become polluted, regardless of its size or location. This includes lakes from remote areas or huge water bodies and is due to the air transportation of pollutant particles and their transfer into precipitation water. The groundwater and surface water consist of swimming pools, ponds, lakes, creeks, rivers, seas, and oceans that may all become polluted at some point. Due to the quick diffusion and dissipation of contamination and the faster natural degradation processes, the bigger the water body is, the shorter the time required for naturally cleansing the pollution and recovery.

Types of Water Pollution

There are various types of water pollution based on the various causes of water pollution. Various classifications can be made, based on various water pollution causes:

1. The type of the water pollutants: based on this classification criteria, water pollution can be:

   i) Chemical: when various chemicals are the water pollution causes. The following chemicals are the most common water pollutants:

   • Crude oil and various petroleum products (including gasoline, diesel fuel, kerosene, motor and lubricating oils, jet fuel). These compounds are lighter than water and thus always sit on top of water forming sheens of "free product". However, part of these compounds dissolve in water and, even in small amounts may be harmful and at the same time may remain unnoticeable by the eye.

   • Fertilizers (including nitrates and phosphates): while small amounts are useful to life, higher amounts of nitrates and phosphates in water are only beneficial to algae and harmful microorganisms and are poisonous to human and aquatic life. These contaminants cannot be seen themselves in water (as they do not form sheens or color the water), but their effects can. The typical effect of water pollution by fertilizers (usually through agricultural runoff) is the fast and abundant water growth.

   • Chlorinated solvents (including TCE, PCE, 1,1,1-TCA, carbon tetrachloride, Freons) which sink in water (are denser than water) and are quite persistent and toxic. These compounds thus, cannot be seen by the eye, in contrast with petroleum products that are easily seen as sheens on top of water surface.

   • Petroleum solvents (including benzene, toluene, xylenes, ethylbenzene).

   • Other organic solvents and chemicals (such as acetone, methyl ethyl ketone, alcohols such as ethanol, isopropanol; or oxygenate compounds such as MTBE).

   • Antibiotics and other pharmaceutical products.

   • Perchlorate: perchlorate salts are used in rocket fuels, as well as many other applications such as fireworks, explosives, road flares, inflation bags, etc. This contaminant is usually associated with military bases, construction sites (when explosives are used). However, natural formation in arid areas may account for perchlorate in water, too (e.g., in Chile, Texas or California where natural formation of perchlorate has been observed).

   • Trihalomethanes: these are usually byproducts of water chlorination and may pollute groundwater and surface water via leaking sewer lines and discharges. Examples of such compounds are: chloroform, bromoform, dichlorobromo methane.

   • Metals and their compounds: of higher health risk are the organo-metal compounds which may form when metals from water react with organic compounds from water. Common examples include Hg, As, and Cr poisoning of water. Thus, if water is polluted with both metals and organic compounds the health risk is higher. And so is the effect of water pollution on aquatic life.

- Pesticides/insecticides/herbicides: comprise a large number of individual chemicals that get into water due to agricultural activities directly (by spraying over large areas) or indirectly with agriculture runoff. The insecticide DDT is a typical example of such type of water pollutant.

- PCBs: in spite of their recent ban, their ubiquitous environmental presence makes these contaminants usually associated with urban runoffs.

ii) Radiological: when radioactive materials are the water pollutant causes.

iii) Biological: when various microorganisms (e.g., bacterial species and viruses), worms, and/or algae occurring in a large number are the water pollution causes. This type of pollution is caused by decaying organic material in water, animal wastes, as well as improper disposal of human wastes.

2. The type (grouping) of the source of water pollutants:

i.) Point Sources: These are localized sources like an industrial process, a mining activity, etc. These sources are usually regulated so that the effect may be predicted and the impact minimized. However, accidental leaks and spills are an exception to that.

ii.) Non-Point Sources: These are unidentified sources from which pollutants are carried away by water discharges and runoffs. Non-point pollution may involve a broad range of pollutants, but in lower amounts than the point sources.

Sources of Water Pollution

The main sources of pollution are all resulted from the disposal of chemical substances coming from medical, industrial and household waste, chaotic agricultural fertilizers disposal and accidental oil spills that pollute the water to a large extent.

Examples of major water pollutants that affect the health of humans are:

- The numerous infectious agents (bacteria, viruses, and parasites) that contaminate the water through sewage, human waste, and animal excreta.

- Radioactive waste that contains highly toxic materials such as uranium, thorium, and radon. This waste is a major water pollutant resulted from mining activities, power plants or natural sources.

- The chemical substances that contaminate the water. These chemicals can be either organic - pesticides, plastic, oil, detergents, etc. - coming from domestic, industrial or agricultural waste, or inorganic - acids, metals, salts - domestic and industrial effluents.

Examples of major water pollutants that affect the ecosystem only are the following:

- Plant nutrients like phosphates and nitrates that form various chemical fertilizers, sewage, and manure.

- Oxygen-demanding manures and agricultural waste resulted from sewage and agricultural run-offs.

- Sediments in the soil (silt) following soil erosion, and heated waters used in several industries and power plants.

Affect of Water Pollution

Water pollution may cause a large variety of diseases and poses a serious problem for human health. This is mainly because we may get exposed to polluted water in various ways, including, but not necessarily limited to:

- Drinking polluted water.

- Bathing or showering in polluted water.

- Swimming in polluted water.

- Breathing the vapors of a polluted water while sitting next to a polluted water source.

- Consuming polluted food (meat and/or vegetables) affected by polluted water.

Consuming meat from animals fed with polluted water of food affected by polluted water (e.g. vegetables irrigated with polluted water or grown in an area with polluted groundwater).

Diseases

The effects of water pollution may appear immediately after exposure and be more or less violent in the case of drinking water with a high amount of pollutants. On the other hand, the effects may appear sometime after repetitive exposure to water contaminated with lower amounts of pollutants. The health effects of drinking contaminated water may range from simple intoxication and stomach aches to deadly diseases or sudden death.

Air Pollution

Air pollution is a form of pollution that refers to the contamination of the air, irrespective of indoors or outside. A physical, biological or chemical alteration to the air in the atmosphere can be termed as pollution. It occurs when any harmful gases, dust, smoke enters into the atmosphere and makes it difficult for plants, animals and humans to survive as the air becomes dirty.

Air pollution can further be classified into two sections- Visible air pollution and invisible air pollution. Another way of looking at Air pollution could be any substance that holds the potential to hinder the atmosphere or the wellbeing of the living beings surviving in it. The sustainment of all things living is due to a combination of gases that collectively form the atmosphere; the imbalance caused by the increase or decrease of the percentage of these gases can be harmful for survival.

The Ozone layer considered crucial for the existence of the ecosystems on the planet is depleting due to increased pollution. Global warming, a direct result of the increased imbalance of gases in the atmosphere has come to be known as the biggest threat and challenge that the contemporary world has to overcome in a bid for survival.

Types of Pollutants

In order to understand the causes of Air pollution, several divisions can be made. Primarily air pollutants can be caused by primary sources or secondary sources. The pollutants that are a direct result of the process can be called primary pollutants. A classic example of a primary pollutant would be the sulfur-dioxide emitted from factories

Secondary pollutants are the ones that are caused by the inter mingling and reactions of primary pollutants. Smog created by the interactions of several primary pollutants is known to be as secondary pollutant.

Causes of Air pollution

1. Burning of Fossil Fuels: Sulfur dioxide emitted from the combustion of fossil fuels like coal, petroleum and other factory combustibles is one the major cause of air pollution. Pollution emitting from vehicles including trucks, jeeps, cars, trains, airplanes cause immense amount of pollution. We rely on them to fulfill our daily basic needs of transportation. But, there overuse is killing our environment as dangerous gases are polluting the environment. Carbon Monoxide caused by improper or incomplete combustion and generally emitted from vehicles is another major pollutant along with Nitrogen Oxides that is produced from both natural and manmade processes.

2. Agricultural activities: Ammonia is a very common by product from agriculture related activities and is one of the most hazardous gases in the atmosphere. Use of insecticides, pesticides and fertilizers in agricultural activities has grown quite a lot. They emit harmful chemicals into the air and can also cause water pollution.

3. Exhaust from factories and industries: Manufacturing industries release large amount of carbon monoxide, hydrocarbons, organic compounds, and chemicals into the air thereby depleting the quality of air. Manufacturing industries can be found at every corner of the earth and there is no area that has not been affected by it. Petroleum refineries also release hydrocarbons and various other chemicals that pollute the air and also cause land pollution.

4. Mining operations: Mining is a process wherein minerals below the earth are extracted using large equipments. During the process dust and chemicals are released in the air causing massive air pollution. This is one of the reasons which are responsible for the deteriorating health conditions of workers and nearby residents.

5. Indoor air pollution: Household cleaning products, painting supplies emit toxic chemicals in the air and cause air pollution. Have you ever noticed that once you paint walls of your house, it creates some sort of smell which makes it literally impossible for you to breathe.

Suspended particulate matter popular by its acronym SPM, is another cause of pollution. Referring to the particles afloat in the air, SPM is usually caused by dust, combustion etc.

Effects of Air Pollution

1. Respiratory and heart problems: The effects of Air pollution are alarming. They are known to create several respiratory and heart conditions along with Cancer, among other threats to the body. Several millions are known to have died due to direct or indirect effects of Air pollution. Children in areas exposed to air pollutants are said to commonly suffer from pneumonia and asthma.

2. Global warming: Another direct effect is the immediate alterations that the world is witnessing due to Global warming. With increased temperatures worldwide, increase in sea levels and melting of ice from colder regions and icebergs, displacement and loss of habitat have already signaled an impending disaster if actions for preservation and normalization aren't undertaken soon.

3. Acid Rain: Harmful gases like nitrogen oxides and sulfur oxides are released into the atmosphere during the burning of fossil fuels. When it rains, the water droplets combines with these air pollutants, becomes acidic and then falls on the ground in the form of acid rain. Acid rain can cause great damage to human, animals and crops.

4. Eutrophication: Eutrophication is a condition where high amount of nitrogen present in some pollutants gets developed on sea's surface and turns itself into algae and adversely affect fish, plants and animal species. The green colored algae that is present on lakes and ponds is due to presence of this chemical only.

5. Effect on Wildlife: Just like humans, animals also face some devastating effects of air pollution. Toxic chemicals present in the air can force wildlife species to move to new place and change their habitat. The toxic pollutants deposit over the surface of the water and can also affect sea animals.

6. Depletion of Ozone layer: Ozone exists in earth's stratosphere and is responsible for protecting humans from harmful ultraviolet (UV) rays. Earth's ozone layer is depleting due to the presence of chlorofluorocarbons, hydro chlorofluorocarbons in the atmosphere. As ozone layer will go thin, it will emit harmful rays back on earth and can cause skin and eye related problems. UV rays also have the capability to affect crops.

There are two types of sources that we will take a look at: Natural sources and Man-made sources.

Natural sources of pollution include dust carried by the Natural sources of pollution include dust carried by the wind from locations with very little or no green cover, gases released from the body processes of living beings (Carbon dioxide from humans during respiration, Methane from cattle during digestion, Oxygen from plants during Photosynthesis). Smoke from the combustion of various inflammable objects, volcanic eruptions etc. along with the emission of polluted gases also make it to the list of Natural sources of Pollution.

While looking at the man-made contributions towards air pollution, smoke again features as a prominent component. The smoke emitted from various forms of combustion like in bio mass, factories, vehicles, furnaces etc. Waste used to create landfills generate methane, that is harmful in several ways. The reactions of certain gases and chemicals also form harmful fumes that can be dangerous to the wellbeing of living creatures.

Solutions for Air Pollution

1. Use public mode of transportation: Encourage people to use more and more public modes of transportation to reduce pollution. Also, try to make use of car pooling. If you and your colleagues come from the same locality and have same timings you can explore this option to save energy and money.

2. Conserve energy: Switch off fans and lights when you are going out. Large amount of fossil fuels are burnt to produce electricity. You can save the environment from degradation by reducing the amount of fossil fuels to be burned.

3. Understand the concept of Reduce, Reuse and Recycle: Do not throw away items that are of no use to you. In-fact reuse them for some other purpose. For e.g. you can use old jars to store cereals or pulses.

4. Emphasis on clean energy resources: Clean energy technologies like solar, wind and geothermal are on high these days. Governments of various countries have been providing grants to consumers who are interested in installing solar panels for their home. This will go a long way to curb air pollution.

5. Use energy efficient devices: CFL lights consume less electricity as against their counterparts. They live longer, consume less electricity, lower electricity bills and also help you to reduce pollution by consuming less energy.

Noise Pollution

Sound is essential to our daily lives, but noise is not. Noise is generally used as an unwanted sound, or sound which produces unpleasant effects and discomfort on the ears.

Sound becomes unwanted when it either interferes with normal activities such as sleeping, conversation, or disrupts or diminishes one's quality of life. Not all noise can be called noise pollution. If it does not happen regularly, it may be termed as 'Nuisance'

Scientists also believe that its not only humans who are affected. For example, water animals are subjected to noise by submarines and big ships on the ocean, and chain-saw operations by timber companies also create extreme noise to animals in the forests.

Generally, noise is produced by household gadgets, big trucks, vehicles and motorbikes on the road, jet planes and a helicopter hovering over cites, loud speakers etc.

Noise (or sound) is measured in the units of decibels and is denoted by the dB. Noise which is more than 115 dB is tolerant. The industrial limit of sound in the industries must be 75 dB according to the World Health Organization.

Noise is considered as environmental pollution, even though it is thought to have less damage on humans than water, air or land pollution. But people who are affected by severe noise pollution know that it is a massive issue that needs attention.

Sources Of Noise Pollution

Noise can come from many places. Let us see a few good sources:

Household Sources

Gadgets like food mixer, grinder, vacuum cleaner, washing machine and dryer, cooler, air conditioners, can be very noisy and injurious to health. Others include loud speakers of sound systems and TVs, iPods and ear phones. Another example may be your neighbor's dog barking all night every day at every shadow it sees, disturbing everyone else in the apartment.

Social Events

Places of worship, discos and gigs, parties and other social events also create a lot of noise for the people living in that area. In many market areas, people sell with loud speakers, others shout out offers and try to get customers to buy their goods. It is important to note that whey these events are not often, they can be called 'Nuisance' rather than noise pollution.

Commercial and Industrial Activities

Printing presses, manufacturing industries, construction sites, contribute to noise pollutions in large cities. In many industries, it is a requirement that people always wear earplugs to minimize their exposure to heavy noise. People who work with lawn mowers, tractors and noisy equipment are also required to wear noise-proof gadgets.

Transportation

Think of aero planes flying over houses close to busy airports like Heathrow (London) or Ohare (Chicago), over ground and underground trains, vehicles on road—these are constantly making a lot of noise and people always struggle to cope with them.

Effects of Noise Pollution

Generally, problems caused by noise pollution include stress related illnesses, speech interference, hearing loss, sleep disruption, and lost productivity. Most importantly, there are three major effects we can look at:

Hearing

The immediate and acute effect of noise pollution to a person, over a period of time, is impairment of hearing. Prolonged exposure to impulsive noise to a person will damage their eardrum, which may result in a permanent hearing impairment.

Marine Animals

Marine scientists are concerned about excessive noise used by oil drills, submarines and other vessels on and inside the ocean. Many marine animals, especially whales, use hearing to find food, communicate, defend and survive in the ocean. Excessive noises are causing a lot of injuries and deaths to whales. For example, the effect of a navy submarine's sonar can be felt 300 miles away from the source.

(SONAR is the use of sound by submarines and other fishing vessels to deterring the depth of water, the closeness of an object, or detects movement of other objects in the water)

This is not only about whales, but the larger marine life is all affected in one way or the other.

Effects on General Health

Health effects of noise include anxiety and stress reaction and in extreme cases fright. The physiological manifestations are headaches, irritability and nervousness, feeling of fatigue and decreases work efficiency. For example, being pounded by the siren of fire fighters, police or ambulance in your city all night everyday leave people (especially elderly people) stresses and tired in the morning.

It is worth noting that these effects may not sound troubling, but the truth is, with time, the consequences can be very worrying.

Noise Pollution Prevention and Control Tips

Below are a few things people and governments can do to make our communities and living laces quieter:

- Construction of soundproof rooms for noisy machines in industrial and manufacturing installations must be encouraged. This is also important for residential building—noisy machines should be installed far from sleeping and living rooms, like in a basement or garage.

- Use of horns with jarring sounds motorbikes with damaged exhaust pipes, noisy trucks to be banned.

- Noise producing industries, airports, bus and transport terminals and railway stations to sighted far from where living places.

- Community law enforcers should check the misuse of loudspeakers, worshipers, outdoor parties and discos, as well as public announcements systems.

- Community laws must silence zones near schools / colleges, hospitals etc.

- Vegetation (trees) along roads and in residential areas is a good way to reduce noise pollution as they absorb sound.

Soil Pollution

Soil pollution is when there are changes in soil caused by the adding or dumping of harmful, unwanted materials, which are called pollutants.

In the case of contaminants, which occur naturally in soil, even when their levels are not high enough to pose a risk, soil pollution is still said to occur if the levels of the contaminants in soil exceed the levels that should naturally be present.

Some examples of pollutants include:

- Trash, including plastic bags and bottles.

- Debris like cement and bricks.

- Metal from old cars or building materials.

- Hospital trash such as needles and bandages.

- Toxic chemicals like batteries, paints, insecticides, fertilizers and other chemicals.

Pollutants into the Soil

- Leakages in sewage systems, underground storage tanks and leaching of soluble substances from landfills can also result in contamination. Rainwater or floods from other polluted lands and water bodies spread contaminants to soils in other locations.

- Farmers often use fertilizers that have chemicals in them to make food grow bigger and faster.

- Improper disposal of paint, chemicals in batteries, household cleaners, car oils, gasoline and so forth can cause soil pollution.

- Oil spills can put oil directly in the soil and in water that seeps into the soil.

- Car accidents can cause oils and gasses to seep into the ground.

Causes of Soil Pollution

There are two main causes through which soil pollution is generated: anthropogenic (man-made) causes and natural causes.

Natural Causes: Natural processes can lead to an accumulation of toxic chemicals in the soil. This type of contamination has only been recorded in a few cases, such as the accumulation of higher levels of perchlorate in soil from the Atacama Desert in Chile, a type of accumulation that is purely due to natural processes in arid environments.

One example of natural causes leading to soil pollution is Acid Rains. This phenomenon occurs when the pollutants present in the air interact with the rain water and fall on land. The polluted water from such rainfall mixes with the groundwater thus making it unfit for human consumption.

Man-made: The Man-made contaminants are the main causes of soil pollution and consist of a large variety of contaminants or chemicals, both organic and inorganic. They can pollute the soil either alone or combined with several natural soil contaminants. Man-made soil pollution is usually caused by the improper disposal of waste coming from industrial or urban sources, industrial activities, and agricultural pesticides. Unlike the waste produced by nature i.e. dead leaves, rotten vegetables, and fruits, animal carcasses, etc., which only adds to the fertility of the soil, the human waste is full of harmful toxins and chemicals that cannot be broken down by the forces of nature. Therefore, man-made waste is the primary reason behind the contamination of soil and the various ill-effects attached to it.

- Industrial Engagement: Increased industrial activity has emerged as the main culprit leading to soil pollution. Most industries rely on raw materials in the form of minerals that are extracted from the earth by way of mining. In the event of the byproducts being polluted and then not disposed-off in the prescribed manner, the resulting waste from various industries remains in the soil and renders it useless or infertile.

- Waste Management: One of the pressing needs of our times is how we manage our waste disposal systems. Apart from the industrial waste mentioned above, every living being contributes to pollution by way of producing bodily waste in the form of urine and the body's solid waste. Most of this waste finds its way in the sewer system and the landfills. When the human waste enters the landfills it contaminates the water and the soil with the toxins and chemicals present in the human bodies.

- Agricultural Engagement: With the increase in population there's also a need to grow more food to meet our requirements. With the advent of technology came the use of pesticides and fertilizers to aid the agricultural process. These pesticides and fertilizers are loaded with strong, harmful, and unnatural chemicals that are not biodegradable. These harmful chemicals enter the soil and render it infertile and also harm the soil composition and make it vulnerable to erosion by air or water. Vegetable and plants that are grown on such land also retain these chemicals and cause pollution of the soil when they decompose.

- Oil-Spills: Every now and then we keep hearing about accidents during storage or transportation of fuel leading to huge amounts of oil being spilled onto the roads or in our oceans. Not only do these oil spills harm the marine life but they also enter our groundwater through the soil thereby making it unfit for human consumption. Accidents involving oil spills at fuel stations or storage facilities lead to the oil entering the soil. The harmful chemicals present in the fuel damage the quality of soil thus making it unfit for future cultivation.

Types of Soil Pollutants

Soil pollution consists of pollutants and contaminants. The main pollutants of the soil are the biological agents and some of the human activities. Soil contaminants are all products of soil pollutants that contaminate the soil. Human activities that pollute the soil range from agricultural practices that infest the crops with pesticide chemicals to urban or industrial wastes or radioactive emissions that contaminate the soil with various toxic substances:

- Biological Agents.
- Agricultural Practices.
- Radioactive Pollutants.
- Urban Waste.
- Industrial Waste.

Examples of Soil Contaminants

There is a large variety of pollutants that could poison the soil. Examples of the most common and problematic soil pollutants can be found below:

- Lead (Pb).
- Mercury (Hg).
- Arsenic (As).
- Copper (Cu).
- Zinc (Zn).
- Nickel (Ni).
- Pahs (Polyaromatic Hydrocarbons).
- Herbicides/Insecticides

Effects of Soil Pollution

Some of the major ill-effects of soil pollution can be highlighted as under:

- Human & livestock health: It's a well-established fact that the human survival is dependent on the soil and a polluted or contaminated soil can only pose serious health hazards to humans. The vegetables or crops cultivated on polluted soil transfer the contamination to the human bodies consuming them. A sustained interaction with such contaminated soil can even change our genetic framework. Many minor and congenital diseases can be attributed to soil pollution and it can even riddle our bodies with many long-term illnesses which require sustained treatment. Soil pollution is also harmful to the livestock feeding on it and can cause food- poisoning with grave outcomes. Soil pollution can also lead to scarcity of food if the soil is rendered infertile and crops fail to grow on it thereby leading to acute food shortage or famines. Long term exposure to the chemicals present in contaminated soil can even lead to deadly diseases like leukemia, cancer of the liver, Central Nervous System Disorders, muscular blockages, etc.

- Effects on agriculture: Contamination can drastically change the ecological balance of the soil within a short period thereby making the crops vulnerable. This can lead to low fertility, making the soil unfit for agriculture. Soil pollution can make large tracts of land barren or dangerous to our health. It can also destroy the fungi and bacteria that hold the soil together and make the soil susceptible to erosion.

- Effects on air & water: The polluted soil releases harmful compounds into the atmosphere and the toxic chemicals present in such soil can enter and contaminate the groundwater which reaches rivers, lakes, oceans, or streams.

- Effects on the fertility of soil: Contaminated soil carrying harmful chemicals adversely affect the soil fertility thereby decreasing the final output of crop. The crop cultivated on polluted soil is unfit for human consumption and can cause serious health problems like food poisoning, etc.

- Harmful fumes: The polluted landfills emit toxic/foul gases and pollute the surroundings and pose serious health hazards to the people living in the nearby areas.

- Effects on soil composition: The harmful chemicals leading to soil pollution also cause the destruction of colonies of various organisms inhabiting the soil (e.g. the earthworms, etc.). This leads to an imbalance in the composition of the soil structure.

Effects on Living Things

Soil pollution can have a number of harmful effects on the ecosystem, humans, plants and animal health. The harmful effects of soil pollution may come from direct contact with polluted soil or from contact with other resources, such as water or food which has been grown on or come in direct contact with the polluted soil.

While anyone is susceptible to soil pollution, soil pollution effects may vary based on age. General health status and other factors, such as the type of pollutant or contaminant inhaled or ingested also matter. However, children are usually more susceptible to exposure to contaminants, because they come in close contact with the soil by playing on the ground. Combined with lower thresholds for disease, this triggers higher risks in them than adults.

Regular exposure to benzene is known to cause leukemia in both children and adults and exposure to polychlorinated biphenyls (PCBs) is linked to liver cancer. Soil pollution can also cause neuromuscular blockage as well as depression of the central nervous system, headaches, nausea, fatigue, eye irritation, and skin rash.

Soil does not need to be highly contaminated to be harmful to humans. Soil that is not significantly polluted may still harm humans directly through bioaccumulation, which occurs when plants are grown in lightly polluted soil, which continuously absorbs molecules of the pollutants. Since the plants cannot get rid of these molecules, they get accumulated over a period of time, causing higher amounts of pollution to exist in the plant than in the soil. Animals that eat many of these polluted plants take on all the pollution those plants have accumulated. Larger animals who eat the plant-eating animals take on all the pollution from the animals they eat. Humans who eat plants or animals that have accumulated large amounts of soil pollutants may be poisoned, even if the soil itself does not contain enough pollution to harm human health.

Prevention of Soil Pollution

Here are few ways to reduce soil pollution:

Reduce Deforestation and Begin Reforestation: Deforestation and soil erosion are very much interconnected. For example, the effects of acid rain and floods can decimate healthy soil in the absence of trees, which would otherwise help absorb and maintain these waters and the toxins that come along. Soil erosion can occur when there are no trees or few plants to prevent the top layer of soil from being removed and transported by forces of nature, such as water and air, which contribute to soil pollution. Through reforestation efforts and planting new vegetation in areas that are prone to erosion, soil pollution can be further prevented.

Mycoremediation or using fungus to metabolize contaminants and accumulate heavy metals.

Avoid Intensive Farming Practices such as over-cropping and over-grazing, as it leads to flood and soil erosion and further deterioration of the soil layer.

Reduce Your 'Waste Footprint': Waste such as plastic, non-biodegradable materials, and litter can accumulate in fertile land, polluting and altering the chemical and biological properties of soil. According to the Clean Air Council, almost one-third of the waste in the U.S. comes from packaging. So, try to purchase materials with the least amount of packaging and always Reduce, Reuse and Recycle.

Phytoremediation or using plants (such as willow) to extract heavy metals.

Discover Soil Washing that uses water to remove contaminants from soil by "scrubbing" it to remove and separate the portion that is the most polluted. Soil washing reduces the amount of soil needing further cleanup and is typically used along with other methods to clean up the soil as it is usually not sufficient enough to do the job alone. It allows the clean-up of polluted soil at one place without having to excavate.

Remediation of oil contaminated sediments with self-collapsing air micro bubbles.

Discover Bioremediation: Use and incite the growth of naturally-occurring microorganisms to break down contaminants and remediate soil pollution by using them as a food source during the aerobic processes. It requires the right temperature, nutrients, and amount of oxygen in the soil.

Use Soil Additives such as lime and organic matter from composting, which can adjust soil pH to sustainable levels and reduce soil erosion and pollution.

Cleanup Options

Cleanup or environmental remediation is analyzed by environmental scientists who utilize field measurement of soil chemicals and also apply computer models (GIS in Environmental Contamination) for analyzing transport and fate of soil chemicals. Various technologies have been developed for remediation of oil-contaminated soil and sediments. There are several principal strategies for remediation:

- Excavate soil and take it to a disposal site away from ready pathways for human or sensitive ecosystem contact. This technique also applies to dredging of bay muds containing toxins.

- Aeration of soils at the contaminated site (with attendant risk of creating air pollution).

- Thermal remediation by introduction of heat to raise subsurface temperatures sufficiently high to volatize chemical contaminants out of the soil for vapour extraction. Technologies include ISTD, electrical resistance heating (ERH), and ET-DSP.

- Bioremediation, involving microbial digestion of certain organic chemicals. Techniques used in bioremediation include land farming, bio stimulation and bioaugmentating soil biota with commercially available micro flora.

- Extraction of groundwater or soil vapor with an active electromechanical system, with subsequent stripping of the contaminants from the extract.

- Containment of the soil contaminants (such as by capping or paving over in place).

- Phytoremediation, or using plants (such as willow) to extract heavy metals.

- Mycoremediation or using fungus to metabolize contaminants and accumulate heavy metals.

- Remediation of oil contaminated sediments with self-collapsing air microbubbles.

Agricultural Pollution

Agriculture is a source of economic development and livelihood on one hand, but pollution due to it can lead to a number of environmental and health hazards. The nature of pollutants and the way they behave in environment are of high importance. Agricultural   pollution is defined as the phenomena of damage, contamination and degradation of environment and ecosystem, and health hazards due to the by- products of farming practices. The relationship of agriculture with the biotic and biotic factors of environment forms a loop known as PSR loop:

- Pressure (P): stress on environment from agricultural activities causes changes in the state or condition of environment.

- State (S): condition of the present environment and its resources.

- Response (R): as shown by the society to the stresses on the changing environ-mental conditions.

There is a need for reliable information about our environment, composition and properties of variety of agricultural pollutants, and their mode of action to under-stand pollution hazards that resulted due to agriculture.

Pollution by agricultural practices has come up ever since the demand for food has increased, proportional to the increase in population. To increase the yield of farms and fields the farmers have had to resort to additional chemical fertilizers, pesticides, weedicides, hormonal treatments for the animals, nutrient laden feed and many such practices which changed the way farming was done traditionally.

Agricultural pollution is contamination of the environment and related surroundings as a result of using the natural and chemical products for farming. This contamination is actually injurious to all living organisms that depend on the food on cultivation.

Causes of Agricultural Pollution

*   Chemical fertilizers

    These are mostly nitrogen and phosphorus based chemicals like ammonia and nitrates that in correct amounts boost the fertility of the soil. But in most cases these are used in more quantity than required and hence tend to be retained in the soil not adding to its goodness.

*   Chemical pesticides

    When pests and insects cause losses on a large scale, this leads to economic fallout for the farmers. Pesticides and insecticides like organ chlorines, organophosphates and carbonates are toxic to the pests. They also tend to bio accumulate i.e. they collect in the body of the organism and lead to chronic poisoning. This can be passed up the food chain. Some pesticides also are absorbed naturally by the plants themselves and stored their different parts. Pesticides are not discriminatory in nature as they also cause harm to beneficial insects such as bees and pollinators.

*   Heavy metals

    Cadmium, fluoride, radioactive elements like uranium are regularly found in the parent minerals from which the fertilizers are obtained. Dangerous metals such as Mercury, Lead, Arsenic, Chromium, and Nickel are seen in traces in Zinc rich wastes from the steel industries which are used as fertilizers. These are often not removed from the because of the high cost involved.

- Excessive tillage of the land

  Overturning, digging or stirring leads to release of greenhouse gases produced in the ground such as nitrous oxide.

- Soil erosion

  Loss of soil material due to poor management causes soil to become infertile.

- Soil sedimentation

  The soil or sediments carried off into water bodies cause a lot of harm. Sedimentation reduces the transportation capability of navigation channels. It reduces the amount of sunlight reaching the water beds affecting the plants and animals living in it. The turbidity it causes interferes with the feeding patterns of the fishes and affects their population. Sedimentation also affects the transport and accumulation of water pollutants.

- Introduction of foreign species

  Many instances of foreign species of plants, animals and insects were introduced to control pests and weeds. But after a while these have taken over and become nuisances themselves. They cause harm to indigenous flora and fauna competing for the natural resources, and also cause changes in the bio diversity. There has been loss of many indigenous beneficial creatures due to this kind of biological pest control.

- Genetic Modification to increase resistance to pest and diseases

  A raging topic of debate today, it is a cause of concern for many that these crops will lead to the loss of many original species and may become weeds themselves. If these will be toxic to consumers ranging from insects to humans is to be studied in depth.

- Animal management

  Farms specializing in rearing cattle, sheep, goats, pigs, and poultry must have strict regulations concerning the disposal of manure and other associated waste material. These must not be indiscriminately disposed in the surrounding areas. They cause pollution of the air as well as the water. 18 per cent of Greenhouse gases are said to be generated by farm animals. The large amounts of manure created, carry pathogens that are harmful for humans too. Proper animal waste management can reduce the huge bulk of it, making it easier to use.

Mechanisms: Types of Agricultural Pollution

Leaching and Ground Water Poisoning

When chemicals accumulate in the soil, depending on its water solubility and soil structure it percolates through reaching the ground water, causing its contamination. This also depends on the rainfall. For example after applying pesticides on crops in sandy areas, if excessive irrigation is done, the pesticide chemicals leach into ground. Leaching occurs not only in the fields, but also at the manufacturing, mixing and disposal sites.

Water Runoff

Only a fraction of fertilizers and other chemical additives are utilized on the fields. The major bulk mixes in the runoff water and flows into the nearby watercourses. This is mainly in the form of nitrates and phosphates.

Eutrophication

Eutrophication is also called water nutrient enrichment. When chemicals rich in nitrogen and phosphorus are flushed into water from farms, this increases the nutrient levels in the water. This encourages the growth of the plants.

Algae benefit greatly from this. When they flourish and bloom they use up most of the oxygen levels in the water body leaving very little for the other life forms in it. This leads to the death of the fish and other animals in the water that cannot swim away to safety.

Algae blooms restrict the penetration of sunlight into the water. This affects the photosynthesis in plants and does not allow the restoration of oxygen levels by this process. Ultimately the water becomes unfit to support any form of life.

Certain toxins are also released by the algae themselves which travel up the food chain and affect the higher life forms adversely.

Types of Impacts by Agriculture on Water Systems

The assessment of various impacts of agriculture on water systems is not easy because the relationship between agriculture and its impact on water bodies is quite complicated as described earlier. Generally various agricultural activities like application of chemical fertilizers, livestock and poultry breeding, aquaculture, and rural population are responsible for increased chemical oxygen demand (COD), ammonia–N, nitrogen, and phosphorus levels which are released into the receiving water systems.

Impacts on Surface Water Quality

Agriculture raises various problems like waterlogging, erosion, salinization, and desertifi- cation. In addition to this, it may also bring up the issues of water quality degradation by agrochemicals, salts, and toxic leachates. In recent years, the salinization of water bodies is a more widespread phenomenon and possibly has greater concern than that of soil salinization. In the past few years, some trace elements like selenium, molybdenum, and arsenic have gathered attention, and their presence in agricultural drainage causes pollution that poses a threat for the progression of few irrigation plans.

Impact on Enrichment of Water

Excessive application of nitrogen- and phosphorus-based fertilizers on agricultural lands of Europe as well as other developed countries has led to enrichment of nitro-gen and phosphorous in surface water, groundwater, and soil. This process of leakage of these nutrients into water bodies ultimately causes eutrophication, and according to one report, about 55 % of excessively enriched surface water in Europe is caused by agriculture mostly by the loss of these nutrients from soil into surface water. Similarly, such type of pollution contributed by agriculture is recognized as major factor for the poor water quality in the USA. A study was conducted in Shanghai, China, to analyze the concentration of nitrogen and phosphorous in groundwater and surface water. These results showed that this nonpoint source-polluted ground-water and surface water was unfit for drinking purposes. The concentration of nitro-gen in surface water was 6.3 mg per liter, while in ground-water; 16.85 mg of nitrogen per liter was calculated. Further studies confirmed that pollution of groundwater is caused by nitrogen-based fertilizer used in peach orchard.

Farmers apply these nitrogen- and phosphorus-based fertilizers to increase the pro-duction or output of the crops. These two nutrients are required for the growth of the plant; however, plants use only the required amounts of these nutrients, and excess nutrients are usually associated with leaching and excessive runoff transfer from land to water bodies. The damage done by such nutrient losses to the water bodies would be very much site specific and would depend on various factors like unstable interaction of water systems, type of soil, chemistry of atmosphere, and fertilizer practices on farm level. Pig farming has emerged rapidly in the province of Hungyen, Vietnam, few years back. With this widespread trend of pig farming, agricultural efficiency flourished in Vietnam economically and was officially encouraged by the Vietnamese government. Four types of such pig farming systems run in the country, i.e., (1) VAC system, which is formed by combining fruit cultivation and fish farming with pig farming; (2) AC system, which is a joint system of fi sh and pig farming; (3) VC system, a blend of pig farming and fruit cultivation; and (4) C system, only comprising of pig farming.

However recently it was indicated that such farming systems became a potential source of pollution of surface and groundwaters in the Red River Delta and its vicinities. The excreta of pig are a supplier of sufficient amounts of nitrogen, phosphate, and potash (0.5 %, 0.3 %, and 0.4 %, respectively). The surface water is dangerously polluted following the contravention of recommended water quality in each of the farming system. The parameters like DO (dissolved oxygen), BOD (bio-logical oxygen demand), COD (chemical oxygen demand), and ammonium and phosphate concentrations exceeded the set values by Vietnamese water standards. Similarly exceeding level of ammonium that breached the standards set by national technical regulation of Vietnam contaminates groundwater. Overall the water pollution was highest in C and VC systems followed by relatively low pollution in VAC and AC systems.

Impacts on Public Health

Agriculture has major impact on water quality; thus, the polluted water causes vari-ous water-borne diseases. According to the World Health Organization (WHO), about four million children die each year due to diarrhea, a waterborne disease. The bacteria-causing disease called coliform is excreted by human which gets mixed in drinking water through poor water management. Surface runoff and nonpoint source pollution may consequently contribute towards high levels of pathogens in surface as well as groundwater bodies . According to another WHO report, nitrogen level in groundwater has drastically increased due to excessive farming. Reiff in 1987 made some important discussions about agricultural impact on water quality. He described that due to reservoir construction for irrigation and hydroelectric power production purposes, there is strong evidence of increase in malaria (particularly in Latin America) and schistosomiasis infecting two hundred million individuals in 70 tropics and subtropics. Farmers and children that bathe in infected water are at more risk. Contamination by other nonpoint source pollution may also pose serious health threats.

Microbial contamination of food crops results from either using polluted water for irrigation or its direct contact with the food. Diseases that are commonly associated include typhoid, cholera, ascariasis (caused by Ascaris lumbricoides), amoebiasis (amoeba, Entamoeba histolytica), giardiasis (protozoan, Giardia labia), and enteroinvasive Escherichia coli. Mostly these diseases are caused by consuming ground crops such as cabbage, carrots, or strawberries. Other complications include hormonal disturbances in human, animals, and fi sh. Due to immense importance of endocrine and its secretions in early days of development, toxicological effects of polluted water badly affect reproductive system.

Prevention of Agricultural Pollution

The priority is to keep the nitrogen and phosphorus rich nutrients from running off into the water sources near fields and animal farms.

- Prevention can never be a solo effort. The state governments, farmers' organizations, collectives and cooperatives, educational institutions and conservation groups need to work together for regulating and reducing farming related water pollution.

- Planning the application of fertilizer at the right time, in the right quantity with the correct methods can reduce the run off.

- Planting certain grasses and clovers that can absorb and recycle the additional nutrients and prevent soil erosion. Planting rows of trees and shrubs around fields and along the borders of the stream or lake also help in the same way.

- Over tilling of the soil must be avoided to prevent soil erosion and soil compaction.

- Managing the correct disposal of animal wastes and keeping farm animals away from water will reduce the nitrogen pollution of the water.

- Composting, solid liquid separation, anaerobic digestion and lagoons are different ways of managing animal manure. Of these anaerobic digestion is the most effective. It involves the use of anaerobic bacteria and heat. The products of this process are nutrient rich liquid used as fertilizer and methane gas that can be burned to produce electricity and heat. Anaerobic digestion is a best method for controlling odor associated with manure management.

Plastic Pollution

The world population is living, working, vacationing, increasingly conglomerating along the coasts, and standing on the front row of the greatest, most unprecedented, plastic waste tide ever faced.

Washed out on our coasts in obvious and clearly visible form, the plastic pollution spectacle blatantly unveiling on our beaches is only the prelude of the greater story that unfolded further away in the world's oceans, yet mostly originating from where we stand: the land.

For more than 50 years, global production and consumption of plastics have continued to rise. An estimated 299 million tons of plastics were produced in 2013, representing a 4 percent increase over 2012, and confirming and upward trend over the past years. In 2008, our global plastic consumption worldwide has been estimated at 260 million tons, and, according to a 2012 report by Global Industry Analysts, plastic consumption is to reach 297.5 million tons by the end of 2015.

Plastic is versatile, lightweight, flexible, moisture resistant, strong, and relatively inexpensive. Those are the attractive qualities that lead us, around the world, to such a voracious appetite and over-consumption of plastic goods. However, durable and very slow to degrade, plastic materials that are used in the production of so many products all, ultimately, become waste with staying power. Our tremendous attraction to plastic, coupled with an undeniable behavioral propensity of increasingly over-consuming, discarding, littering and thus polluting, has become a combination of lethal nature.

Although inhabited and remote, South Sentinel island is covered with plastic. Plastic pollution and marine debris, South Sentinel Island, Bay of Bengal.

A simple walk on any beach, anywhere, and the plastic waste spectacle is present. All over the world the statistics are ever growing, staggeringly. Tons of plastic debris (which by definition are waste that can vary in size from large containers, fishing nets to microscopic plastic pellets or even particles) is discarded every year, everywhere, polluting lands, rivers, coasts, beaches, and oceans.

Lying halfway between Asia and North America, north of the Hawaiian archipelago, and surrounded by water for thousands of miles on all sides, the Midway Atoll is about as remote as a place can get. However, Midways' isolation has not spared it from the great plastic tide either, receiving massive quantities of plastic debris, shot out from the North Pacific circular motion of currents (gyre). Midways' beaches, covered with large debris and millions of plastic particles in place of the sand, are suffocating, envenomed by the slow plastic poison continuously washing ashore.

Then, on shore, the spectacle becomes even more poignant, as thousands of bird corpses rest on these beaches, piles of colorful plastic remaining where there stomachs had been. In some cases, the skeleton had entirely biodegraded; yet the stomach-size plastic piles are still present, intact. Witnesses have watched in horror seabirds choosing plastic pieces, red, pink, brown and blue, because of their similarity to their own food. It is estimated that of the 1.5 million Laysan Albatrosses which inhabit Midway, all of them have plastic in their digestive system; for one third of the chicks, the plastic blockage is deadly, coining Midway Atoll as "albatross graveyards" by five media artists.

From the whale, sea lions, and birds to the microscopic organisms called zooplankton, plastic has been, and is, greatly affecting marine life on shore and off shore. In a 2006 report, Plastic Debris in the World's Oceans, Greenpeace stated that at least 267 different animal species are known to have suffered from entanglement and ingestion of plastic debris. According to the National Oceanographic and Atmospheric Administration, plastic debris kills an estimated 100,000 marine mammals annually, as well as millions of birds and fishes.

The United Nations Joint Group of Experts on the Scientific Aspects of Marine Pollution (GESAMP), estimated that land-based sources account for up to 80 percent of the world's marine pollution, 60 to 95 percent of the waste being plastics debris.

However, most of the littered plastic waste worldwide ultimately ends up at sea. Swirled by currents, plastic litter accumulates over time at the center of major ocean vortices forming "garbage patches", i.e. larges masses of ever-accumulating floating debris fields across the seas. The most well-known of these "garbage patches" is the Great North Pacific Garbage Patch, discovered and brought to media and public attention in 1997 by Captain Charles Moore. Yet some others large garbage patches are highly expected to be discovered elsewhere.

The plastic waste tide we are faced with is not only obvious for us to clearly see washed up on shore or bobbing at sea. Most disconcertingly, the overwhelming amount and mass of marine plastic debris is beyond visual, made of microscopic range fragmented plastic debris that cannot be just scooped out of the ocean.

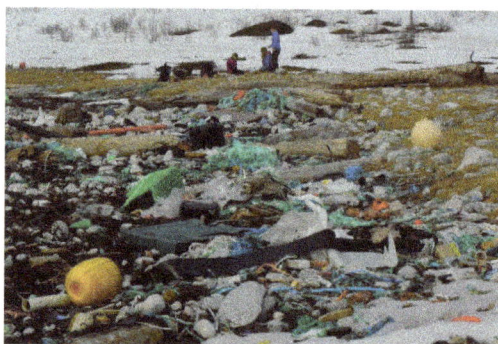

Slow, silent, omnipresent, ever increasing, more toxic than previously thought, the plastic pollution's reality bears sobering consequences, as recently unveiled by the report of Japanese chemist Katsuhiko Saido at the 238th National Meeting of the American Chemical Society (ACS) in August 2009 and the findings from the Project Kaisei and Scripps (Seaplex) scientific cruise-expeditions collecting seawater samples from the Great Garbage Patch. Both, the reports and expeditions uncovered new evidence of how vast and "surprisingly" (as it was termed at the ACS meeting) toxic the plastic presence in the marine environment is.

Environmentalists have long denounced plastic as a long-lasting pollutant that does not fully break down, in other terms, not biodegradable. In 2004, a study lead by Dr Richard Thompson at the University of Plymouth, UK, reported finding great amount of plastic particles on beaches and waters in Europe, the Americas, Australia, Africa and Antarctica. They reported that small plastic pellets called "mermaids tears", which are the result of industry and domestic plastic waste, have indeed spread across the world's seas. Some plastic pellets had fragmented to particles thinner than the diameter of a human hair. But while some cannot be seen, those pieces are still there and are still plastic. They are not absorbed into the natural system, they just float around within it, and ultimately are ingested by marine animals and zooplankton (Plankton that consists of tiny animals, such as rotifers, copepods, and krill, larger animals eggs and larvae's and of microorganisms once classified as animals, such as dinoflagellates and other protozoans.). This plastic micro-pollution, with its inherent toxicity and consequences on the food chain, had yet to be studied.

Dr Saido's study was the first one to look at what actually happens over the years to these tons of plastic waste floating in the world's oceans. The study presents an alarming fact: these tons of plastic waste reputed to be virtually indestructible, do decompose with surprising speed, at much lower temperature than previously thought possible, and release toxic substances into the seawater, namely bisphenol A (BPA) and PS oligomer. These chemicals are considered toxic and can be metabolized subsequent to ingestion, leading Dr Saido to state "plastics in the ocean will certainly give rise to new sources of global contaminations that will persist long into the future".

This past August a different study, from a group of oceanography students from Scripps Institution of Oceanography (SIO), UCSD, accompanied by the international organization Project Kaisei's team, embarked on two vessels, New Horizon and Kaisei, through the North Pacific Ocean to sample plastic debris and garbage. SIO director Tony Haymet described the trip as "as forage into the great plastic garbage patch in the north." To summarize the scientific data collected on the ship, Miriam Goldstein, chief scientist on New Horizon, stated: "We did find debris coming up in our nets in over 100 consecutive net tows over a distance of 1,700 miles. It is pretty shocking." She said, "There is not a big island, not garbage dumps that we can really see easily." She described it more as a place where large debris floats by a ship only occasionally, but a lot of tiny pieces of plastic exist below the surface of the water. "Ocean pretty much looks like ocean," she said. "The plastic fragments are mostly less than a quarter inch long and are below the surface. It took at first a magnifying-glass to see the true extent of plastic damage in the North Pacific."

The overwhelmingly largest unquantifiable plastic mass is just made of confetti-like fragmented pieces of plastic.

All sea creatures, from the largest to the microscopic organisms, are, at one point or another, swallowing the seawater soup instilled with toxic chemicals from plastic decomposition. The world population " is eating fish that have eaten other fish, which have eaten toxin-saturated plastics. In essence, humans are eating their own waste."

The scientists from Project Kaisei and Scripps hope their data gives clues as to the density and extent of these debris, especially since the Great Pacific Garbage Patch might have company in the Southern Hemisphere, where scientists say the gyre is four times bigger." We're afraid at what we're going to find in the South Gyre, but we've got to go there," said Tony Haymet.

The "Silent World" is shedding mermaid tears. A plastic-poison has undeniably been instilled by us, prompting an unwilling and illegitimate confrontation of two titans: one synthetic (plastic), the other oceanic. The crisis is of massive proportion. An unprecedented plastic tide has occurred, pervasively affecting the world's oceans, beaches, coasts, seafloor, animals and ultimately, us.

The Great Plastic Tide: Magnitude, Scope, Extent

A full understanding of the magnitude and scope of this plastic pollution starts with clear definitions as to what and why it is happening. Thus, we will define the notions of marine debris, gyres, and oceanic garbage patches, or giant floating marine debris field, as first discovered in the North Pacific by Captain Charles Moore's, since referred to as The Great Pacific Garbage Patch (GGP).

Marine Debris and Plastic

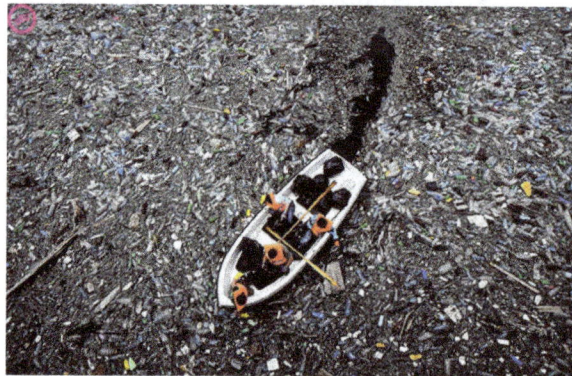

Marine Debris

The term marine debris has been used for at least 25 years to refer to man-made materials that have been discarded or lost into the ocean. The earliest references come from the 1984 Workshop on the Impacts and Fate of Marine Debris. This workshop came out of a 1982 request from the Marine Mammal Commission to the National Marine Fisheries Service to examine the impacts of marine debris. At that time, the focus of research was primarily on derelict fishing gear.

Other terms used prior to 1984 include the following: man-made debris, synthetic debris, plastic litter, floating plastic debris, man-made objects, and debris.

Mouth of the Los Angeles River, Long Beach, California.

The term marine debris encompasses more than plastic, including metals (derelict vessels, dumped vehicles, beverage containers), glass (light bulbs, beverage containers, older fishing floats), and other materials (rubber, textiles, lumber). Plastic certainly makes up the majority of floating litter, but in some areas the debris on the ocean floor may contain sizeable amounts of those other denser types.

Scientists have similarly and more simply defined marine debris as, any manufactured or processed solid waste material that enters the ocean environment from any source. Marine debris is definitely characterized as human-created waste that has deliberately or accidentally become afloat. They tend to accumulate at the center of gyres and on coastlines, frequently washing aground where it is known as beach litter.

The US Congress passed a bill in 2006, The Marine Debris Research, Prevention, and Reduction Act, to create a program to address the marine debris pollution. One of the requirements in the bill was for NOAA (National Oceanic and Atmospheric Administration) and the U.S. Coast Guard, to promulgate a definition of marine debris for the purposes of the Act. Thus, USCG and NOAA drafted and published a definition of marine debris in September 2009. The definition is this: "Any persistent solid material that is manufactured or processed and directly or indirectly, intentionally or unintentionally, disposed of or abandoned into the marine environment or the Great Lakes." Marine debris can come in many forms, from a plastic soda bottle to a derelict vessel. Types and components of marine debris include plastics, glass, metal, Styrofoam, rubber, derelict fishing gear, and derelict vessels.

UNEP has defined marine debris, or marine litter, as "any persistent, manufactured, processed, or solid material discarded, disposed of, or abandoned in the marine and coastal environment." This is an even more global and comprehensive definition, as it does include the marine and correlated coastal impact of the aforementioned litter.

As we mentioned supra, land-based sources of debris account for up to 80 percent of the world's marine pollution. Such debris is unquestionably one of the world's most pervasive pollution problems affecting our beaches, coasts, oceans, seafloors, inland waterways and lands. It affects the economies and inhabitants of coastal and waterside communities worldwide. The effect of coastal littering is obviously compounded by vectors, such as rivers and storm drains, discharging litter from inland urban areas. Obviously, ocean current patterns, climate and tides, and proximity to urban centers, industrial and recreational areas, shipping lanes, and commercial fishing grounds influence the types and amount of debris that is found in the open ocean or collected along beaches, coasts and waterways, above and below the water's edge.

The other 20 percent of this debris is from dumping activities on the water, including vessels (from small power and sailboats to large transport ships carrying people and goods), offshore drilling rigs and platforms, and fishing piers.

Over the past 60 years, organic materials, once the most common form of debris, have yielded to synthetic elements as the most abundant material in solid waste. Marine litter is now 60 to 80 percent plastic, reaching 95 percent in some areas, according to a report by the Algalita Marine Research Foundation, published in October 2008 in Environmental Research.

Around and around, worldwide, at distant seas, or merely bobbing among the waves before washing up ultimately on shore, a daily and ever too common plastic spectacle is unveiled: bottles, plastic bags, fishnets, clothing, lighters, tires, polystyrene, containers, plastics shoes, just a myriad of man-made items, all sharing a common origin: us.

Yearly data adds to the despondent reality of how extensively the plastic tide is increasingly affecting world's beaches and coasts. Launched in 1986 by the Ocean Conservancy, the Center for Marine Conservation's annual International Coastal Cleanup (ICC) has grown into the world's largest volunteer effort to collect data on the marine environment. Held the third Saturday of each September, the International Coastal Cleanup engages the public to remove trash and debris from the coasts, beaches, waterways, underwater, and on lands to identify the sources of debris. It is a compelling global snapshot of marine debris collected on one day at thousands of sites all over the world. The 2008, 23rd ICC reported that 104 countries and locations, from Bahrain to Bangladesh, and in 42 US States, from southern California to the rocky coast of Maine, had participated. The overwhelming percentage of debris collected was plastics and smoking paraphernalia. The 2008 report states that plastic litter has increased by 126 percent since ICC first survey in 1994. The top 3 items found in 2008 were cigarettes butts, plastic bags, and food wrappers/containers.

Durable and slow to degrade, plastic materials that are used in the production of so many products, from containers for beverage bottles, packing straps and tarps, and synthetic nylon materials used in fishing line, all become debris with staying power. Plastics debris accumulates because it does not biodegrade as many other substances do; although it will photo degrade on exposure to sunlight and does decompose, more rapidly than previously thought.

In addition, most of these plastic waste items are highly buoyant, allowing them to travel in currents for thousands of miles, endangering marine ecosystems and wildlife along the way. Marine debris is a global trans boundary pollution problem.

The instillation of plastic in an oceanic world vests a terrible reality. Because of the properties of plastic as a synthetic material and because of the absence of boundary, vastness, currents and winds at seas, this resilient polluting material is being spread worldwide by an even more powerful vehicle, the seas. It appears then daunting, impossible, a priori, to control, efficiently clean-up, remedy effectively, even sufficiently study the plastic pollution. This unwilling confrontation of titans, one plastic the other oceanic, has become ineluctably a crisis of massive proportion.

Plastic

The paucity of concerted and definitive scientific data/research in this matter is staggering compared to the extent of the problem.

Only in 1997, with Captain Charles Moore's discovery, was the plastic waste pollution in the ocean widely brought to media light and finally began to receive more serious attention from the public and the scientific world, stepping the way to more exhaustive research about plastic and its consequences and effects when entering marine life.

Of the 260 million tons of plastic the world produces each year, about 10 percent ends up in the Ocean, according to a Greenpeace report (Plastic Debris in the World's Oceans). Seventy percent of the mass eventually sinks, damaging life on the seabed. The rest floats in open seas, often ending up in gyres, circular motion of currents, forming conglomerations of swirling plastic trash called garbage patches, or ultimately ending up washed ashore on someone's beach.

But the washed up or floating plastic pollution is a lot more than an eyesore or a choking/entanglement hazard for marine animals or birds. Once plastic debris enters the water, it becomes one of the most pervasive problems because of plastic's inherent properties: buoyancy, durability (slow photo degradation), propensity to absorb waterborne pollutants, its ability to get fragmented in microscopic pieces, and more importantly, its proven possibility to decompose, leaching toxic Bisphenol A (BPA) and other toxins in the seawater.

"Plastics are a contaminant that goes beyond the visual", says Bill Henry of the Long Marine Laboratory, UCSC.

But before we develop further the realities and consequences of the plastic-covered beaches, seafloor and plastic-instilled seawater, it is necessary to present simple facts about plastic itself.

Facts About Plastic

A simple definition could be: any of a group of synthetic or natural organic materials that may be shaped when soft and then hardened, including many types of resins, resinoids, polymers, cellulose derivatives, casein materials, and proteins: used in place of other materials, as glass, wood, and metals, in construction and decoration, for making many articles, as coatings, and, drawn into filaments, for weaving. They are often known by trademark names, as Bakelite, Vinylite, or Lucite.

In chemistry, plastics are large molecules, called polymers, composed of repeated segments, called monomers, with carbon backbones. A polymer is simply a very large molecule made up of many smaller units joined together, generally end to end, to create a long chain. The smallest building block of a polymer is called a monomer. Polymers are divided into two distinct groups: thermoplastics (moldable) and thermosets (not). The word "plastics" generally applies to the synthetic products of chemistry.

Alexander Parkes created the first man-made plastic and publicly demonstrated it at the 1862 Great International Exhibition in London. The material, called parkesine, was an organic material derived from cellulose that, once heated, could be molded and retained its shape when cooled.

Many, but not all, plastic products have a number – the resin identification code – molded, formed or imprinted in or on the container, often on the bottom. This system of coding was developed in 1988 by the U.S.-based Society of the Plastics Industry to facilitate the recycling of post-consumer plastics. It is indeed, quite interesting to go through the fine lines.

1. Polyethylene terephthalate (PET or PETE): Used in soft drink, juice, water, beer, mouthwash, peanut butter, salad dressing, detergent, and cleaner containers. Leaches antimony trioxide and (2ethylhexyl) phthalate (DEHP).

2. DEHP is an endocrine disruptor that mimics the female hormone estrogen. It has been strongly linked to asthma and allergies in children. It may cause certain types of cancer and it has been linked to negative effects on the liver, kidney, spleen, bone formation, and body weight. In Europe, DEHP has been banned since 1999 from use in plastic toys for children under the age of three.

3. High-density polyethylene (HDPE): Used in opaque milk, water, and juice containers, bleach, detergent and shampoo bottles, garbage bags, yogurt and margarine tubs, and cereal box liners. Considered a safer plastic. Research on risks associated with this type of plastic is ongoing.

4. Polyvinyl chloride (V or Vinyl or PVC): Used in toys, clear food and non-food packaging (e.g., cling wrap), some squeeze bottles, shampoo bottles, cooking oil and peanut butter jars, detergent and window cleaner bottles, shower curtains, medical tubing, and numerous construction products (e.g., pipes, siding). PVC has been described as one of the most hazardous consumer products ever created. Leaches di (2-ethylhexyl) phthalate (DEHP) or butyl benzyl phthalate (BBzP), depending on which is used as the plasticizer or softener (usually DEHP). DEHP and BBzP are endocrine disruptors mimicking the female hormone estrogen; have been strongly linked to asthma and allergic symptoms in children; may cause certain types of cancer; and linked to negative effects on the liver, kidney, spleen, bone formation, and body weight. In Europe, DEHP, BBzP, and other dangerous phthalates have been banned from use in plastic toys for children under three since 1999. Not so elsewhere, including Canada and the United States.

   Dioxins are unintentionally, but unavoidably, produced during the manufacture of materials containing chlorine, including PVC and other chlorinated plastic feedstocks. Dioxin is a known human carcinogen and the most potent synthetic carcinogen ever tested in laboratory animals. A characterization by the National Institute of Standards and Technology of cancer causing potential evaluated dioxin as over 10,000 times more potent than the next highest chemical (diethanol amine), half a million times more than arsenic, and a million or more times greater than all others.

5. Low-density polyethylene (LDPE): Used in grocery store, dry cleaning, bread and frozen food bags, most plastic wraps, and squeezable bottles (honey, mustard). Considered a safer plastic. Research on risks associated with this type of plastic is ongoing.

6. Polypropylene (PP): Used in ketchup bottles, yogurt and margarine tubs, medicine and syrup bottles, straws, and Rubbermaid and other opaque plastic containers, including baby bottles. Considered a safer plastic. Research on risks associated with this type of plastic is ongoing.

7. Polystyrene (PS): Used in Styrofoam containers, egg cartons, disposable cups and bowls, take-out food containers, plastic cutlery, and compact disc cases. Leaches styrene, an endocrine disruptor mimicking the female hormone estrogen, and thus has the potential to cause reproductive and developmental problems. Long-term exposure by workers has shown brain and nervous system effects and adverse effects on red blood cells, liver, kidneys, and stomach in animal studies. Also present in secondhand cigarette smoke, off gassing of building materials, car exhaust, and possibly drinking water. Styrene migrates significantly from polystyrene containers into the container's contents when oily foods are heated in such containers.

8. Other: This is a catchall category that includes anything that does not come within the other six categories. As such, one must be careful in interpreting this category because it includes polycarbonate – a dangerous plastic – but it also includes the new, safer,

biodegradable bio-based plastics made from renewable resources such as corn and potato starch and sugar cane. Polycarbonate is used in many plastic baby bottles, clear plastic sippy cups, sports water bottles, three and five gallon large water storage containers, metal food can liners, some juice and ketchup containers, compact discs, cell phones, and computers. Polycarbonate leaches Bisphenol A (some effects described above) and numerous studies have indicated a wide array of possible adverse effects from low-level exposure to Bisphenol A: chromosome damage in female ovaries, decreased sperm production in males, early onset of puberty, various behavioral changes, altered immune function, and sex reversal in frogs.

Plastic debris, of all sizes and shapes, is a transboundary pollution problem with a powerful vehicle, the ocean.

Buoyancy

Plastics travel long distances. Their distribution in the oceans isn't uniform, yet they are omnipresent from the Polar Regions to the Equator. Scientists are still refining methods to detect and analyze the materials. A good example of plastic debris' buoyancy is as follows. In 1992, twenty containers full of rubber ducks were lost overboard from a ship traveling from China to Seattle. By 1994, some had been tracked to Alaska, while others reached Iceland in 2000. The ducks (with a distinctive logo on their base) have been sighted in the Arctic, Pacific and Atlantic Oceans.

Photodegradation vs. Biodegradation

Plastic is generally a durable material. Its durability has made the culprit of the problem since it is considered resistant to natural biodegradation processes, i.e. the microbes that break down other substances do not recognize plastic as food. Yet plastic can be fragmented with the effects of UV, being broken down by light in smaller and smaller debris over time.

Biodegradation, the breaking down of organic substances by natural means, happens all the time in nature. All plant-based, animal-based, or natural mineral-based substances will over time biodegrade. In its natural state raw crude oil will biodegrade, but man-made petrochemical compounds made from oil, such as plastic, will not. Why not? Because plastic is a combination of elements extracted from crude oil then re-mixed up by men in white coats. Because these combinations are manmade they are unknown to nature. Consequently, it has been thought that there is no natural system to break them down. The enzymes and the microorganisms responsible for breaking down organic materials that occur naturally such as plants, dead animals, rocks and minerals, don't recognize them. This means that plastic products are said indestructible, in a biodegradable sense at least.

In sum, as time passes, we know that plastic will eventually photo-degrade, i.e. break down into smaller and smaller fragments by exposure to the sun. The photo-degradation process continues down to the molecular level, yet photo-degraded plastic remains a polymer. No matter how small the pieces, they are still and always will be plastic, i.e. they are not absorbed into or changed by natural processes. At sea, the plastic fragmentation process occurs as well, due to wave, sand action, and oxidation. Estimates for plastic degradation at sea have been ranged from 450 to 1,000 years.

Of particular concern are the floating small plastic fragments often referred in the media to as mermaids' tears, which are tiny nurdles of raw plastic resin that form the building material of every manufactured plastic product, or are granules of domestic waste that have fragmented over the years. Dr Richard Thompson of the University of Plymouth, UK has identified plastic particles thinner than the diameter of a human hair. But while they cannot be seen, those pieces are still there and are still plastic. Not absorbed into the natural system, they just float around within it. He estimates that there are 100,000 particles of plastic per sq km of seabed and 300,000 items of plastic per sq km of sea surface.

Either way, mermaid tears, or fragmented plastic debris, reaching microscopic size over time, remain everywhere and are almost impossible to clean up. They are light enough to float in the wind, landing in the earth's oceans. Mermaid's tears are often found in filter feeders like mussels, barnacle, lugworm and amphipods.

Thus, the photo degradation of plastic debris makes the matter worse. Plastic becomes microscopic, invisible, yet ever polluting waters, beaches, coasts, seafloor, being eaten by even tinier marine organisms, therefore entering the food chain insidiously and ineluctably.

Toxic Sponge

Corroborating reports and findings worldwide demonstrated that fragmented plastics debris' increase and massive presence on and off shores does constitute reason for raised worries and awareness.

Studies on small plastic pellet by Dr Richard Thompson and by Hideshige Takada, Yukie Mato professor of organic geochemistry at Tokyo University, have shown that plastic debris meeting other pollutants in the oceans absorbs harmful chemicals from the sea water they float in, acting like a pollution sponges.

These studies have been conducted on plastic nurdles not just because of their uniform size and shape, thus easier to study and compare by scientifics, but also because of their wide spread presence on the world's beaches.

In UK, mermaid tears are the second common plastic litter found on the beaches according to the Marine Conservation Society's 2007 data and a Surfers against Sewage (SAS) report.

According to Charles Moore, these resin pellets account for around 8 percent of annual oil production and are the raw material for the 260 million tons of plastic consumed yearly worldwide. Lightweight and small, they escape in untold volumes during transport and manufacture and wash in the ocean.

Even though these researches have been conducted on nurdles, it is crucial to keep in mind, as Dr. Takada team confirmed, that other types of plastic debris (from fishing gear, shopping bags, to small fragments) displays the exact same propensity as the nurdles of raw plastic resin to absorb toxins.

Plastic resin pellets are round, shiny and tiny, mostly less than 5mm in diameter. The very structure of the plastic material is oily and greasy (basically plastics are solid oil) therefore promoting the accumulation of hydrophobic contaminants (ones that tend to repel and not absorb water) from the surrounding seawater. Chemicals like PCB's and DDE are very hydrophobic. It was shown that plastic pellets suck up these dangerous persistent organic pollutants (POPs) and toxins with a concentration factor that's almost 1 million times greater compared to the overall concentration of the chemicals in seawater. In other words, waterborne hydrophobic pollutants do collect and magnify on the surface of plastic debris, thus making plastic far more deadly in the ocean than it would be on land.

These findings, published in the Marine Pollution Bulletin, were based on samples gathered from 30 beaches in 17 countries. PCB (Polychlorinated biphenyls) pollutant concentrations on plastic pellet were highest on US coasts, followed by Western Europe and Japan. The highest concentrations of DDT (Dichlorodiphenyltrichloroethane), the most toxic of all pesticides, were found on the US west coast and Vietnam.

Plastic marine debris, thought to be "indestructible", "lasting forever", has been shown to decompose faster than previously thought, under unexpected conditions (in the water and at sea temperature) and, most importantly, releasing toxic substances not found in the natural element: seawater.

Decompose

Since plastics belong to a chemical family of high polymers, they are essentially made up of a long chain of molecules containing repeated units of carbon atoms. Because of this inherent molecular stability (high molecular weight), plastics do not easily breakdown into simpler components.

Plastics do decompose, though not fully, over a very long period of time (in average 100 to 500 years). Commercially available plastics (polyolefins like polyethylene, polypropylene, etc.) have been further made resistant to decomposition by means of additional stabilizers like antioxidants. Thus, unless the plastic is specially designed to decompose in the soil, such materials can last a very long time because the chemical bonds that hold the molecules together are often stronger than nature's power to take them apart. This means that soil microorganisms that can easily attack and decompose things like wood and other formerly living materials cannot break the various kinds of strong bonds that are common to most plastics. This depends upon the plastic (polymer) and the environment to which it is exposed.

The Marine Conservancy has published that the estimated decomposition rates of most plastic debris found on coasts are:

- Foamed plastic cups: 50 years
- Plastic beverage holder: 400 years
- Disposable diapers: 450 year
- Plastic bottle: 450
- Fishing line: 600 years.

Until Dr. Saido's report, no studies had been conducted on plastic decomposition at low temperature in the marine environment, owing to the mistaken conception that plastic does practically not decompose in such condition. In the first study to look at what happens over the years to the billions of pounds of plastic waste drifting in the world's oceans, researchers, lead by Katsuhiko Saido, PhD, reported that plastic does "decompose with surprising speed (as little as a year) and release potentially toxic substances into the water."

These findings were reported on August 19, 2009, at the 238th National Meeting of the American Chemical Society (ACS). The scientists there termed the discovery "surprising."

Dr. Saido described a new method to simulate the breakdown of plastic products at low temperatures (30 °Celsius, 86 °F), such as those found in some oceans. David Barnes, marine ecologist from the British Antarctic Survey, expressed that the Japanese's team lab results cannot be applied uniformly across the ocean. However, even though the decomposition process would not occur in much cooler seawater as Barnes mentioned, the oceans are vast, currents are constant and permanent, nothing stays static and furthermore, it seems that garbage patches where plastics accumulate, are to be found in even greater dimension in the South Gyres, in the tropical and sub-tropical zones with very warm waters. One of the researchers stated: "Even at 30 degrees Celsius, the plastic decomposes. In natural conditions, the tide comes in and sunlight heats the plastics [which increases decomposition."

The type of plastic studied by Saido's team was polystyrene, a white foamed plastic, commonly known by the trademark Styrofoam.

The process involved modeling plastic decomposition at room temperature, removing heat from the plastic and then using a liquid to extract the BPA and PS Oligomer that are not found naturally, thus must have been created through the decomposition of the plastic. Once degraded, the plastic was shown to release three new compounds not found in nature: styrene monomer (SM), styrene dimer (SD) and styrene trimer (ST). While SM is already a known carcinogen, SD and ST are suspected to be as well.

Plastics are not metabolized subsequent to ingestion since they are polymers. On the other hand, low molecular compounds such as PS oligomer or BPA from plastic decomposition are toxic and can be metabolized.

Samples of sea sand and seawater collected from Europe, India, Japan and the Pacific Ocean were found to be contaminated, with up to 150 parts per million of some of these components of plastic decomposition.

"Plastics in daily use are generally assumed to be quite stable," said study lead researcher Katsuhiko Saido, Ph.D. "We found that plastic in the ocean actually decomposes as it is exposed to the rain and sun and other environmental conditions, giving rise to yet another source of global contamination that will continue into the future."

This latest study clearly shows new micro-pollution by compounds generated by plastic decomposition to be taking place out of sight in the ocean, leaching toxic chemicals such as Bisphenol A (BPA) and derivatives of polystyrene.

Even though present in seawater and sands, the pollutants are found in highest concentration in areas heavily littered with plastic debris, such as ocean vortices, which bring us to define more specifically the notion of gyres and "garbage patches".

Gyres and Garbage Patches

The plastic litter defacing the beaches of the World, alarming in Hawaiian archipelagos for instance, led, only two decades ago, a couple of private and public teams of environmentalists and scientists to start conducting research regarding marine debris in the oceans.

Between 1985 and 1988, an Alaska- based team of researchers found high concentrations of marine debris accumulating in regions governed by vortices like pattern of ocean currents, mentioning the high probability of the existence of "a large area highly concentrating plastic waste debris in the North Pacific". Flyovers of the area have been conducted as well, but not in a conclusive way. The trash was not that obvious from the sky. Indeed, despite its size and density, the GGP is not visible from satellite photography because of its consistency, as Kaisei project and Scripps teams confirmed last August. The largest mass of the plastic pollution contains fragmented pieces of plastic, permeating the ocean, almost invisible to the naked eye, suspended at, or beneath the surface of the ocean.

Charles Moore, a Californian sailor, surfer, volunteer environmentalist, and researcher, was crossing the Pacific Ocean while returning from a trans Pacific sailing race in 1997. He veered from the usual sea route taking a shortcut across the edge of the North Pacific Ocean. He came upon an area, the Doldrums, a windless part of the ocean that mariners usually avoid. The area is filled with tiny phytoplankton, but few big fish or mammals, thus fishermen and sailors rarely travel through it. There, Charles Moore saw an ocean he had never known.

The area that Charles Moore came upon, the North Pacific Subtropical Gyre, is a slowly moving, clockwise spiral or vortex of currents created by a high-pressure system of air currents. He reported his find to Curtis Ebbesmeyer, an oceanographer, who named it the Eastern Garbage Patch.

Shocked by the extent of the plastic litter, Charles Moore went on alerting the world to the existence of this phenomenon.

Moore's discovery was finally corroborating previous scientists's, suspicions and extrapolations in regard to the existence of a high debris concentration in stable bodies of oceanic waters created over time by the rotating ring-like ocean currents system called gyres.

"We were out in the middle of the Pacific, where you would think the ocean would be pristine," recalls the Alguita's captain, Charles Moore. "And instead, we get the Exxon Valdez of plastic-bag spills."

Captain Moore's giant floating debris field's discovery has since been subject to other expeditions, and another "patch" was found further west.

Media light was finally brought in force at that point. Human kind has walked on the moon since 1969 yet the ocean was still quite an unknown frontier in our collective conscience.

Gyres

The North Pacific gyre has given birth to two large masses of ever-accumulating plastic debris, known as the Western and Eastern Pacific Garbage Patches, collectively called the Great Pacific Garbage Patch (GGP). It is a gyre of marine litter in the Central North Pacific Ocean stretching for hundreds of miles across the ocean 1,000 miles from California coast on the East, to Japan and Hawaii on the West.

More specifically, a gyre is a large-scale circular feature made up of ocean currents that spiral around a central point, clockwise in the Northern Hemisphere and counterclockwise in the Southern Hemisphere. Gyres make up to 40 percent of the ocean. That is 25 percent of the globe. All of them are accumulators of debris, Moore says.

Worldwide, there are five major subtropical oceanic gyres: the North and South Pacific Subtropical Gyres, the North and South Atlantic Subtropical Gyres, and the Indian Ocean Subtropical Gyre. Since each behaves in the same vortex style, scientists are certain that massive conglomerates of marine litter like the North Pacific Garbage Patch exist in each of the world's oceans. That is soberingly self-explanatory: such huge garbage patch, or even larger ones, is more than likely to be discovered in the near future.

It is very difficult to measure the exact size of a gyre because it is a fluid system, but the North Pacific Subtropical Gyre is roughly estimated to be approximately 7 to 9 million square miles, approximately three times the area of the continental United States (3 million square miles). Gyres do potentially aggregate debris on that large a scale. That is titanic.

Upon returning from their 22 days venture on the GGP, Project Kaisei and Scripps scientists' stated in a press conference held in September 2009: "(we) hope our data gives clues as to the density and extent of marine plastic debris, especially since the Great Pacific Garbage Patch may have company in the Southern Hemisphere, where scientists say the gyre is four times bigger. "We're afraid at what we're going to find in the South Gyre, but we've got to go there," said Tony Haymet, director of the Scripps Institution.

Garbage Patches

The Great Garbage patch is two separate accumulations connected by a 6,000-mile marine litter "corridor" known as the North Pacific Convergence Zone (STCZ). As will be explained infra, the convergence zone is in itself another serious accumulator of traveling plastic debris.

The Eastern Pacific Garbage Patch floats between Japan and Hawaii; the Western Patch floats between Hawaii and California. The rotational pattern created by the North Pacific Gyre draws in waste material from as far as Asia to the USA. As material is captured in the currents, wind-driven surface currents gradually move floating debris inward, trapping debris in higher concentrations in the calm center. Ocean currents carry debris from the East coast of Asia to the center, in less than a year, and from the Western US in about 5 years.

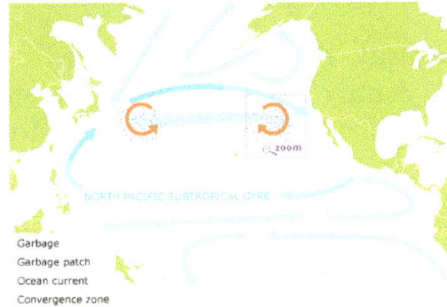

NOAA has tracked the great pacific garbage patch movements to some degree. It is not a stationary area, but one that moves and changes as much as a thousand miles north and south, and during warmer ocean periods, known as el nino, it drifts even further south. The movements occur because the north pacific gyre is made up of four different currents: the north pacific current to the north; the california current to the west; the north equatorial current to the south; and the kuroshio current to the east. This movement sometimes brings the western garbage patch within 500 nautical miles of the california coast and causes extraordinary massive debris pile-ups on beaches, such as in the Hawaiian Islands and Japan.

The name garbage patch has led many to believe that this area is a large and continuous patch of easily visible marine debris items, such as bottles and other litter, akin to a literal blanket of trash that should be visible with satellite or aerial photographs. This is simply not true. While larger litter items can be found in this area, along with other debris such as derelict fishing nets, the largest mass of the debris is small bits of floatable plastic.

We cannot emphasize enough that the GGP is now characterized by extremely high concentrations of suspended plastic debris for 90 percent, basically a soupy mix of plastic-filled seawater, made of tiny plastic debris that have been trapped by the currents and stretching for maybe thousands of miles, and that is the great problem.

Indeed, the researchers from Project Kaisei and the Scripps Environmental Accumulation of Plastic Expedition (SEAPLEX), after their journey through the area, collecting samples the whole way, reported: "All we could see, not at first glance but with magnifying glass and magnifying worries, for miles and miles, was an incredibly huge mass of confetti-like tiny mermaid tears, plastic fragments, floating just beneath the surface." As one of the scientist from the Project Kaisei witnessed: "There's no island, there's no eighth continent" Miriam Goldstein said, "It doesn't look like a garbage dump. It looks like beautiful ocean. But then when you put the nets in the water, you see all the little pieces." And while the expedition covered 1,700 miles, members of the Kaisei team say the patch could be much, much larger.

As for its depth and assumed density, the scientists reported that the GGP's waters were just clogged with plastic particles to a depth of 10 meters below the surface. The Scripps/ Kaisei survey mission of the gyre found that plastic debris was present in 100 consecutive samples taken at varying depths and net sizes.

In sum, they estimated the patch area ranged in size from 700,00 km2 to more than 15 million km2; the area may contain over 100 million tons of plastic debris.

Already in 1999, a study by Charles Moore, sampling waters from the GGP, found that the concentrations of plastic there reached one million particles per square mile, topping the concentration of zooplankton (plankton consisting of small animals and the immature stages of larger animals) by a factor of six. In 2008, the published new research from the Algalita foundation team of scientists estimated that the number had doubled.

After the return of the two vessels from Project Kasei and Scripps (Seaplex), Kaisei and New Horizon, the only certainty was that the size of the Great Pacific Garbage Patch remains uncertain. "It's not a hard and fast number." There has been extensive media coverage about the garbage patch over the past couple years; however, its reported size and mass have differed from in different reports. "It's a little bit like a whirlpool on the surface of a river or a lake. You'd be hard-pressed to tell me where the edge is. All you know is that it's stronger in the middle than it is in the outer reaches. But it's an area of many hundreds of miles, perhaps thousands, in which the ocean currents tend to bring it together," according to Robert Knox, deputy director for research at the Scripps Institution.

In the summer of 2010, Project Kaisei will launch its second expedition to the North Pacific Gyre where it will send multiple vessels to continue marine debris research and, in particular, to test an array of larger marine debris collection systems.

The Eastern Garbage Patch has been studied the most so far, yet it is not and obviously cannot be the only vast oceanic "rubbish dump out there" says Charles Moore. The GGP is definitely not the only type of area where marine debris concentrates. Several other features within the ocean, including oceanic eddies and convergence zones, can lead to debris accumulation as well. A great and well known example is the North Pacific Subtropical Convergence Zone (STCZ). It is located along the southern edge of an area known as the North Pacific Transition Zone. NOAA has focused on the STCZ because it is an area triggering massive debris accumulation in Hawaii. This area does not have distinct boundaries and varies in location and intensity of convergence throughout the year. This zone moves seasonally between 30° and 42° N latitude (approximately 800 miles), extending farther south (28° N) during periods of El Nino.

The August 2009 expedition was so dauntingly successful and the findings were so "shocking", that "Scripps' officials are now working to raise funds for a trip to the South Pacific gyre, sometime within the next two years", Haymet announced at a press conference last August. "That project's scope is far greater. While the North Pacific patch appeared as large as the Continental US, its South Pacific cousin is suspected to be about four times as large", roughly the size of all of Western and part Eastern Europe.

The Path to Successful Resolution

This unprecedented plastic waste tide appears as vast as the ocean, as ungraspable as the unfathomable mass of microscopic plastic fragments present at sea, transported by winds and currents, yet, ultimately, the plastic tide can become as limited as our chosen relationship with plastics, which involves a dramatic behavioral change on our part. The path to successful resolution of the crisis clearly appears as we are the problem and the solution.

The Victims and the Aggressors

The despondent effects and too numerous casualties of the great plastic tide are visible, but more alarmingly, beyond visual, which ought to prompt the perpetrators to choose no other path than the advocacy and culture of consistent and sustained behavioral changes.

The Victims

Animals

From the whale, sea lions, and birds to the microscopic organisms called zooplankton, plastic has been, and is, greatly affecting marine life, i.e animals on shore and off shore, whether by ingestion or entanglement.

In a 2006 report, Plastic Debris in the World's Oceans, Greenpeace stated that at least 267 different species are known to have suffered from entanglement and ingestion of plastic debris. The National Oceanographic and Atmospheric Administration said that plastic debris kills an estimated 100,000 marine mammals annually, millions of birds and fishes.

The largest pieces of marine plastic debris, miles long discarded fishing nets and lines mostly, take an obvious toll on animals. These derelicts nets, called ghost nets, snare and drown thousands of larger sea creatures per year, such as seals, sea lions, dolphins, sea turtles, sharks, dugons, croco-diles, seabirds, crabs, and other creatures. Acting as designed, these nets restrict movement caus-ing starvation, laceration, infection, and, in animals that need to return to the surface to breathe, suffocation.

On shores, researchers have also watched in horror as hungry turtles wolf down jellyfish-like plas-tic bags and seabirds mistake old lighters and toothbrushes for fish, choking when they try to re-gurgitate the plastic trash for their starving chicks.

In the waters, plastic bags specifically, can be mistaken as food and consumed by a wide range of marine species, especially those that consume jellyfish or squid, which look similar when floating in the water column.

Albatross and others birds are choosing plastic pieces because of their similarity to their own food as well. Captain Moore and his Alguita team did see, above the GGP, albatrosses and tropicbirds circling above the line of trash. With little else to choose, they were obviously eating plastic. The birds seemed to be picking and choosing "the reds and pinks and browns. Anything that looks like shrimp," Moore says. Earlier in the trip, the Alguita had visited the French Frigate Shoals, off Hawaii, home to endangered monk seals and seabird rookeries. In the birds' gullets researchers found red plastic particles. Greenpeace reported that a staggering 80 percent of seabird popula-tions observed worldwide have ingested plastics. Research into the stomach contents of dead Ful-mars from the Netherlands, between 1982 and 2001, found that 96 percent of the birds had plastic fragments in their stomachs with an average of 23 plastic pieces per bird.

When plastic ingestion occurs, it blocks the digestive tract, gets lodged in animals windpipes cut-ting airflow causing suffocation, or fills the stomach, resulting in malnutrition, starvation and po-tentially death. Indeed, it is found that debris often accumulates in the animals' gut and gives a false sense of fullness, causing the animal to stop eating and slowly starve to death.

In April 2002 a dead Minke whale washed up on the Normandy coast in France. An investigation found that its stomach contained 800 kg of plastic bags.

In February 2004, a Cuviers Beaked whale (Ziphius cavirostris) was found washed ashore on the west coast of the Isle of Mull, Scotland. Cuviers beaked whales are rarely seen in coastal waters, as they are predominantly a deep-water species. The Hebridean Whale and Dolphin Trust took various skins and blubber samples and removed the stomach for further study by the Scottish Agricultural College. On initial removal it was found that the entrance to the stomach was completely blocked with a cylinder of tightly packed shredded black plastic bin liner bags and fishing twine. It is believed that this made it difficult for the animal to forage and feed effectively.

50 to 80 percent of sea turtles found dead are known to have ingested plastic marine debris.

The smaller the pieces of plastic get, the more dangerous they are to marine organisms. Fragmented plastic, specifically nurdles and small size mermaid tears, are found in the stomach of smaller sea creatures as well: fish, birds, marine mammal, reptile, jelly fish, select plastic pellets as they resemble fish eggs.

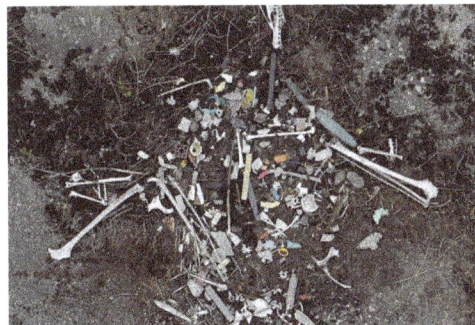

Whether the chemicals contained in the plastics are then desorbed to digestive fluids and transferred to tissues in quantities significant enough to harm the animals is subject to ongoing, yet still incomplete, research. However, as more and more studies on the matter are undergone, unpleasant findings are definitely uncovered.

What is proven, as we've seen supra, is that plastic does soak up pollutants, acting as toxic-sponge for man-made toxins present in the ocean, thus accumulating pollutants such as polychlorinated biphenyls (PCBs) and heavy metals at concentrations up to 1 million times higher than in ocean water. PCBs can lead to reproductive disorders, death, an increased risk of disease, and an alteration of hormone levels. They have been linked to the masculinization of female polar bears and spontaneous abortions and declines in seal populations. In 1988, Ryan et al obtained evidence that PCBs in the tissues of Great Shearwaters were derived from ingested plastic particles. Furthermore, DDT, a pesticide that was banned in the US in the 1960's and labeled by the Environmental Protection Agency in 1987 as a "probable human carcinogen," has been found on these plastics fragments. The most recent review of all evidence concludes that exposure to DDT before puberty increases the risk of breast cancer.

Food Chain

In a September press conference, Doug Woodring from Project Kaisei, said that assessments of the impact of plastic debris on phytoplankton, zooplankton, and mesopelagic (midwater) fishes are undergoing. The samples collected from the seawater will be subject to more scientific studies for the toxicity of the plastics and how this is really affecting our food chain (in ways that are only just becoming known and not good ways).

Katsuhiko Saido, Ph.D said, "We found that plastic in the ocean actually decomposes giving rise to yet another source of global contamination that will continue into the future." Furthermore, as Saido added: "We are concerned that plastic pollution is also caused by these invisible materials and that it will harm marine life." While the potential toxicity of these tiny plastic constituents is still understudied for much of marine life, plastics are abundant in many forms. Plastics, including polystyrene, are common in the wads of accumulated, undigested matter that young black-footed albatrosses cough up before they fledge.

Whether plastics present a unanimously accepted and proven toxic challenge to marine life, and subsequently to humans, is one of the biggest challenges facing scientists right now.

Health

Saido's latest science report last summer about the decomposition of polystyrene plastics vests a simple reality: Bisphenol A (BPA) has been shown and proven to interfere with the reproductive systems of animals. PS oligomer and BPA from plastic decomposition are toxic and can be metabolized, while styrene monomer is a suspected carcinogen. Low levels of BPA and PS oligomer have been proven to cause hormone disruption in animals.

More scientific reports are being published on the effects of Bisphenol A on animal and human health, and the news is not good.

In 2009, a professional, international medical organization in the field of endocrinology and metabolism, The Endocrine Society, reported data from new research on animals experimentally treated with BPA. Studies presented at the group's annual meeting show BPA can affect the hearts of women, can permanently damage the DNA of mice, and appear to be entering the human body from a variety of unknown sources. A 2005 study, which analyzed BPA serum concentrations, concluded that "exposure to BPA is associated with recurrent miscarriage".

The first major study of health effects on humans associated with bisphenol A exposure was published in September 2008 by Iain Lang and colleagues in the Journal of American Association. The cross-sectional study of almost 1,500 people assessed exposure to bisphenol A by looking at levels of the chemical in urine. The authors found that higher bisphenol A levels were significantly associated with heart diseases, diabetes, and abnormally high levels of certain liver enzymes.

A 2008 scientific review concluded that "prenatal exposure to low doses of BPA alters breast development and increases breast cancer risk". A 2009 scientific review, funded by the "Breast Cancer Fund", has recommended "a federal ban on the manufacture, distribution and sale of consumer products containing bisphenol A".

A 2009 study on urinary concentrations concluded that prenatal BPA exposure might be associated with externalizing behaviors in two-year old children, especially among female children.

A 2009 study on Chinese workers in BPA factories found that workers were four times more likely to report erectile dysfunction, reduced sexual desire, and overall dissatisfaction with their sex life than workers in factories that made products ranging from textiles to machinery, in which there was no heightened BPA exposure. They were also more likely to report reduced sexual function within one year of beginning employment at the factory, and the higher the exposure, the more likely they were to have sexual difficulties.

A 2009 review of available studies has concluded, "Prenatal BPA exposure acts to exert persistent effects on body weight and adiposity."

A 2009 scientific review about environmental chemicals and thyroid function concluded, "Available evidence suggests that governing agencies need to regulate the use of thyroid-disrupting chemicals, particularly as such uses relate exposures of pregnant women, neonates and small children to the agents". A 2009 review summarized BPA adverse effects on thyroid hormone action.

All sea creatures, from the largest to the microscopic organisms are, at one point or another, swallowing the seawater soup instilled with toxic chemicals from plastic decomposition. Much of ocean's life is in the microscopic size range and zooplankton is the base of the food chain. As environmentalists remind the world's population, "We are eating fish that have eaten other fish, which have eaten toxin-saturated plastics. In essence, humans are eating their own waste".

Beaches, Coast, Sea Floor, Shorelines

Blatantly visible is the plastic spill washing up on the shores and beaches. Just a walk on any beach, anywhere in the world, and plastic debris are found in one form or another. All over the world the statistics are ever growing, just staggeringly. Last year, an estimated 150,000 tons of marine plastic debris washed up onto the shores of Japan and 300 tons a day on India's shores.

The Hawaiian Archipelago, extending from the southernmost island of Hawaii 1,500 miles northwest to Kure Atoll, is among the longest and most remote island chains in the world. The 19 islands of the archipelago, including Midway atolls, receive massive quantities of plastic debris, shot out from the Pacific gyres. Some of the plastic litter is decades old. Some beaches are buried under 5 to 10 feet of plastic trash, while other beaches are riddled with "plastic sand," millions of grain-like pieces of plastic that are practically impossible to clean up. One of the reasons marine debris accumulates in these islands is the movement of debris within the North Pacific Subtropical Convergence Zone (STCZ).

Two studies on several islands off Jakarta Bay and islands further to the northwest in the Java Sea, reported that debris pollution on shorelines had substantially increased between 1985 and 1995. Both studies noted that results implicated Jakarta as a major source of the debris. On 23 of the islands, it was reported that the total litter at the strandline ranged from not detectable to 29.1 items/m. Plastic bags, polystyrene blocks, and discarded footwear accounted for 80 percent of the items found.

Researchers Barnes and Milner list five studies which have shown increases in accumulation rates of debris on mid to high latitude coasts of the southern hemisphere.

Surveys of shorelines around the world, reported by Greenpeace, have recorded the quantity of marine debris either as the number of items per km of shoreline or the number of items per square meter of shoreline. The highest values reported were for Indonesia (up to 29.1 items per m) and Sicily (up to 231 items per m).

It's been reported by Greenpeace that an estimated 70 percent of the mass of fragmented plastic present in the open oceans of the world does sink to the deep-sea bed. A limited body of literature exists, though, concerning these small to microscopic particles (micro debris) mirroring the little research addressed to marine litter on the sea floor.

Ecosystem Changes

Another effect of the plastic tide that goes beyond visual is its potentiality to change entire ecosystems.

"Plastic is not just an aesthetic problem," says marine biologist David Barnes of the British Antarctic Survey. "It can actually change entire ecosystems." He has documented that plastic debris which floats on the oceans, acts as rafts for small sea creatures to grow and travel on. This represents a potential threat for the marine environment should an alien species become established. It is postulated that the slow speed at which plastic debris crosses oceans makes it an ideal vehicle for this. The organisms have plenty of time to adapt to different water and climatic conditions.

Coral Reefs

Derelict fishing gear can be destructive to coral reefs. Corals are in fact animals, even though they may exhibit some of the characteristics of plants and are often mistaken for rocks. In scientific classification, corals fall under the phylum Cnidaria and the class Anthozoa. They are relatives of jellyfish and anemones.

Nets and lines become snagged on coral and subsequent wave action causes coral heads to break off at points where the debris was attached. Once freed, debris can again snag on more coral and the whole process is repeated. This cycle continues until the debris is removed or becomes weighted down with enough broken coral to sink. Eventually, derelict fishing gear may become incorporated into the reef structure.

Plastic bags can kill coral by covering and suffocating them, or by blocking sunlight needed by the coral to survive. During 2001, so many plastic bags were regularly seen in the Gulf of Aqaba, off the coast of Jordan, that the Board of Aqaba Special Economic Zone issued a law banning the production, distribution, and trade of plastic bags within the areas under their jurisdiction.

Economics

Marine litter cause serious economic losses to various sectors and authorities. Among the most seriously affected are coastal communities (increased expenditures for beach cleaning, public health and waste disposal), tourism (loss of income, bad publicity), shipping (costs associated with fouled propellers, damaged engines, litter removal and waste management in harbors), fishing (reduced and lost catch, damaged nets and other fishing gear, fouled propellers, contamination), fish farming and coastal agriculture.

In 2007, it was written that the costs of river pollution to the economy are enormous. Waterborne diseases are India's leading cause of childhood mortality. Shreekant Gupta, a professor at the Delhi School of Economics who specializes in the environment, estimates that lost productivity from death and disease resulting from river pollution and other environmental damage is equivalent to about 4 percent of gross domestic product.

The bill for cleaning the beaches in Bohuslän, on the west coast of Sweden, in just one year was reportedly at least 10 million SEK or $1,550,200. In Britain, Shetland fishermen reported that 92 per cent of them had recurring problems with debris in nets, with each boat losing between $10,500 and $53,300 per year as a result of marine litter. The cost to the local industry could be as high as $4,300,000. The municipality of Ventanillas in Peru has calculated that it would have to invest around $400,000 a year in order to clean its coastline, while its annual budget for cleaning all public areas is only half that amount.

Our Oceans and coastlines are under unprecedented plastics waste attack. It's coming back at us in many ways. It's a dire problem that only received serious scientific and public attention in the early 90's, as we know, but all along the perpetrators have simply and clearly been identified.

The Aggressors

Behind each and every piece of littered plastic debris there is a human face. At a critical decision point, someone, somewhere, mishandled it, either thoughtlessly or deliberately. Cigarette filters and cigar tips, fishing line, rope and gear, baby diapers and nappies, six-pack rings, beverage bottles and cans, disposable syringes, tires, the litany of plastic litter is as varied as the products available in the global marketplace, but it all shares a common origin.

Sources

260 million tons per year is our estimated plastic consumption, 6 789 billion, is the estimated world population. Our voracious appetite for plastics, coupled with a culture of discarding products that we have chosen for their inherent longevity, is a combination of lethal nature for our environment.

The ultimate symbol of our throwaway lifestyle is the plastic bag: 500 billion to 1 trillion plastic bags is the number consumed annually, which is about a million a minute. The production of plastic bags creates enough solid waste per year to fill the Empire State Building two and a half times. The petroleum used to make only 14 plastic bags could drive a car 1 mile.

Plastic bags are commonly found in waterways, on beaches, and in other unofficial dumping sites across China, for instance. Litter caused by the notorious bags has been referred to as "white pollution."

In the United States, however, measures to ban or curtail the use of plastic bags have met with official resistance. With its powerful lobby, the plastics industry argues that jobs will disappear. The industry employs some two million workers. Americans alone throw out at least 100 billion bags a year, the equivalent of throwing away 12 million gallons of oil, which seems an intolerable waste. Until the U.S. follows the lead of San Francisco, China, Ireland, Uganda, South Africa, Russia, and Hong Kong and targets the reduction of plastic bags using legislature, we each need to make a conscious choice and refuse to use it.

The core of the plastic waste instillation in world's oceans is primarily rooted in poor practices of solid waste management, a lack of infrastructure, various human activities, an inadequate understanding on the part of the public of the potential consequences of their actions, the lack of adequate legal and enforcement systems nationally and internationally, and a lack of financial resources affected to the cause. Mainly a consensus needs to happen, as a culture of behavioral changes needs to be promoted.

The four main land-sources of plastics debris have been identified as:

Shoreline and Recreational Activities Related Litter

This includes: bags, balloons, beverages bottles, cans, caps, lids, shoes, cups, plates, forks, knives, spoons, food wrappers/containers, six-pack holders, pull tabs, shotgun shells/wadding, straws, stirrers, toys, medical hygiene (condom, syringe), drug and smoking paraphernalia (The filters are made of cellulose acetate, a synthetic polymer (fiber) that can last for many years in the environment), and 55 gallons drums. All this land-based debris blows, washes, or is discharged into the water from land areas after people engaged in beach-going activities have discarded it.

About 80 percent of all tourist flock to coastal areas. Massive influxes of tourists, often to a relatively small area, have a huge impact, adding to the pollution of the local population, putting local infrastructure and habitats under enormous pressure. For example, 85 percent of the 1.8 million people who visit Australia's Great Barrier Reef are concentrated in two small areas, Cairns and the Whitsunday Islands, which together have a human population of just 130,000 or so.

Shoreline activities account for 58 percent of the marine litter in the Baltic Sea region and almost half in Japan and the Republic of Korea. In Jordan, recreational activities contribute up to 67 percent of the total discharge of marine litter. This is a particularly big problem in the East Asian Seas region home to 1.8 billion people, 60 percent of whom live in coastal areas with its fast growing shipping and industrial development. Other emerging hotspots include the oil-boom coasts of the Caspian and the littoral states of Iran and Azerbaijan.

In South Asia, the growing ship-breaking industry has become a major source of marine debris. In Gujarat, India one of the largest and busiest ship-breaking yards in the world operations are carried out on a 10-kilometer stretch on the beaches of Alang, generating peeled-off paint chips and other types of non-degradable solid waste making its way into the sea.

Sewage (Waste Waters Containing Plastic Type Products, Rivers, Waterways)

Under normal, dry weather conditions, most wastes are screened out of sewage in countries that do apply strict sewage treatment. However, materials can bypass treatment systems and enter waterways when rain levels exceed sewage treatment facilities' handling capacity. During these times, sewage overflows occur.

The Yamuna River, which flows 855 miles from the Himalayas into the Ganges, is one of India's most, but not only, polluted river. The Centre for Science and Environment says that nearly 80 percent of the river's pollution is the result of sewage. Combined with industrial runoff, that comes to more than three billion liters of waste per day, a quantity well beyond the river's assimilative capacity. Many Indian rivers are so polluted they exceed permissible levels for safe bathing.

The lack of adequate solid waste management facilities results in hazardous wastes entering the waters of the Western Indian Ocean, South Asian Seas, and southern Black Sea, among others.

Fishing Related Debris

Dumping, wastes from ships, boats platforms (20%). Derraik stated that ships are estimated to dump 6.5 million tons of plastic a year. An estimated fourth fifths of the oceanic debris is litter blown seaward from landfills and urban runoff washed down storm drains. Clean up on land where 80 percent of the plastic debris originates is thus the primarily obvious answer.

Manual Clean Up

The simplest, yet highly effective, action is the manual cleanup of the beaches, coasts, rivers, lands and estuaries.

National and international manual clean-up operations of shorelines and sea floor are in existence.

For instance, the past 20 years, the Japan Environmental Action Network (JEAN) has been organizing a yearly beach cleanup and survey.

On an international level, the International Coastal Cleanup (ICC) was installed. The International Coastal Cleanup (ICC) engages the public to remove trash and debris from the world's beaches and waterways, to identify the sources of debris, and to change the behaviors that cause pollution. The origins of the ICC began in 1985 with research conducted by The Ocean Conservancy (then known as the Center for Marine Conservation – CMC) on plastics in the marine environment. Contracted by the U.S. Environmental Protection Agency, Office of Toxic Substances, the CMC produced the report Plastics in the Ocean: More Than a Litter Problem, which was the first study to identify plastics as a significant marine debris hazard. The data collected and analyzed from the annual ICC Cleanup is used locally, nationally and internationally to influence policy decisions, spawn campaigns for recycling programs, support public education programs, launch adopt-a-beach programs, and even storm water system overhaul and legislative reform.

The Clean Up the World program is run in conjunction with UNEP. It engages more than 40 million people from 120 different countries in cleanup operations.

As part of its Rise above Plastics campaign, Surfrider foundation is hosting frequent beach cleanups; it is an example of an encouraging trend towards collective awareness and action to solve the problem at its source.

Worldwide private groups and associations are more and more aware that clean-up does need to happen, one day at a time, one person at a time.

Cleaning up of the Oceans Debris in the Open Seas

NOAA has also been contacted regarding cleanup of the debris directly in the garbage patch and other areas of the North Pacific; however, cleanup is likely to be more difficult than it may seem. "If only things were that simple. We could just go out there and scoop up an island," says Holly Bamford, director of NOAA's marine debris program. "If it was one big mass, it would make our jobs a whole lot easier." It's like a galaxy of garbage, populated by billions of smaller trash islands that may be hidden underwater or spread out over many miles.

Furthermore, in some areas where marine debris concentrates so does marine life, such as in the STCZ. This makes simple scooping up of the material risky, more harm than good may be caused. Straining ocean waters for plastics would capture the plankton that is the base of the marine food web and responsible for 50 percent of the photosynthesis on Earth. (NOAA).

As Captain Charles Moore once said: the cleaning up effort of the oceanic garbage patches "would bankrupt any country and kill wildlife in the nets as it went."

However, confident in the future and investigating new horizons, Doug Woodring, from Project Kaisei, will be producing a documentary for National Geographic testing catch techniques for the plastic waste ("we know not all can be caught, but some can for sure"), at least for the largest debris that we know do decompose over time and actually more rapidly than previously thought.

The cleanup operation is the most immediate, highly effective, and simplest, action/plan that we, the problem, can undertake right now to contribute to the solution. It is a great starting point for a fundamental cultural change that need to occur, which is part of a major consensus.

The Consensus

Undeniably a culture of behavioral changes, now in its infancy, need to further blossom and be implemented/prompted at all levels: individual, associative, governmental, legislative, industrial, technological, educational, philosophical, national, and international.

It simply starts with individual choices. That is the enormous task, yet the enormous power as well because it resides within each and every one of us. Indeed, thanks to an increased awareness of the plastic pollution spread, local, national, individual, and associative actions have taken place worldwide to stop the plastic hemorrhage at the source.

Education, Legislation, and Awareness

Education

The starting point of all greater good does remain education and information.

More and more awareness and preventive programs are promoted.

For instance, in 2004, the Australian government launched a campaign called Keep the Sea Plastic Free, in which it attempted to educate the public to dispose of plastic waste properly.

Surf rider foundation is aiming to raise awareness of plastic marine debris and reduce the proliferation of single-use plastic bags and water bottles, as well as the number one littered item worldwide, cigarette butts. The Rise above Plastics program also seeks to promote a more sustainable lifestyle and educate people about the prevalence of plastic marine debris on our beaches and oceans and how deadly it can be to marine life.

The Indonesian government, for instance "is seriously concerned about improving its waste management and informing the public," quoted the Jakarta Post, 2008. The head of the Maritime and Coastal Resources Studies, Tridoyo Kusumastanto, said that both individual and industrial dumpers should learn from scavengers who take solid waste out instead of dumping it into rivers, canals and the sea. Tridoyo estimated that some 40 tons of waste have been dumped into rivers and other waterways daily in surrounding areas and thus polluting the Java Sea. A campaign against river and sea pollution has been called, and people are urged to change their culture of throwing garbage into waterways and other common places.

Being educated on the situation and aware of the consequences ultimately leads us toward better choices in term of consumption and waste management of plastic at an individual level. It can be as simple as refraining from discarding plastic after first use plastic inherently chosen for its durability.

As H. Takada mentioned: "We can't avoid using plastic, but we use too much. "In fact, he's added a fourth "R" to the ecologist's classic mantra of reduces, reuse, recycle: refuse. The current bring-your-own-bag movement at retail stores and supermarkets is a good start in terms of refusing, he notes.

Instantaneous, prompt eradication of plastics in its current form, rate of production, and consumption is not realistically feasible, yet constant pressure is impacting industry and politicians to "think green," to have environmentally responsible approach, production, prevention plans, and legislations.

Extend Producer Responsibility

Relentless associative campaigns have proven that change can happen, such as the recent victory from the Uk's Surfers Against Sewage (SAS) campaign against mermaid tears.

"SAS launched a campaign to rid British coastlines of mermaid tears, and will continue to build up until factory practice changes." On June 5th 2009, the release of the British Plastic Federation's (BPF) Operation Clean Sweep (OCS) guidance manual was a victory on the preventive field. OCS is aimed at improving British plastic factories efficient use of plastic pellets, commonly referred to as mermaid's tears. SAS initially highlighted the problem of mermaid's tears on UK beaches to the BPF in 2007, delivering a bottle of 10,000 mermaid's tears, collected from one Cornish beach, to a BPF biopolymer seminar. SAS also released a covert film documenting mermaid's tears in the storm drains of plastic factories in the southwest, highlighting the route from factory to beach. SAS and the BPF have worked together on the OCS solution. SAS has already signed up Contico, one of the southwest's largest plastic factories, to pilot some of the improvements within OCS.

Shoichiro Kobayashi, from The Japan Plastics Industry Federation, says that its members have taken measures to reduce spillage of plastics nurdles.

"Awareness of the problem is high," says Kobayashi, and has been since JEAN and other NPOs started publicizing the issue about 15 years ago. The federation has about 1,000 members. Together with the 2,200-member All Japan Plastic Products Industrial Foundation, the two groups represent the largest plastic producing companies in Japan. Kobayashi says his organization encourages members and associated transport companies to avoid spillage and to cover all drainage pipe openings with wire mesh. That's helped reduce the problem at larger companies, but there are more than 20,000 producers of plastic goods in Japan.

On September 22nd 2009 in California, a press conference was held by DTC director Maziar Movassaghi and Project Kaisei founder Mary Crowley, along with representatives from the State of California and various nonprofit groups. They pushed for Extended Producer Responsibility, the philosophy that companies that create products must take responsibility for the full life cycle of those products, products that are "benign by design." Mary Crowley added, "Let's reduce the source of this pollution by not only choosing healthy, plastic-free products ourselves, but also urging our legislators to pass Extended Producer Responsibility legislation. In fact, such a bill is currently on the table in the state of California. AB283, the California Product Stewardship Act, is an important step in this process."

Local legislations, with clear frames and enforcements measures, are increasingly being presented and passed in concert with international programs and legislations, which need ratification by as many countries as possible as the pollution is without frontiers.

Legislation and International Concerted Programs

Internationally

In 1972, the London Convention, a United Nations agreement to control ocean dumping, was entered into. It was followed by the most well-known piece of International legislation, the International Convention for the Prevention of Pollution from ships (MARPOL). MARPOL was introduced in 1988 with the intention of banning the dumping of most garbage and all plastic materials from ships at sea. A total of 122 countries have ratified the treaty. There is some evidence that the implementation of MARPOL has helped to reduce the marine debris problem.

In 1972 and 1974, conventions were held in Oslo and Paris, respectively, which resulted in the passing of the OSPAR Convention, an international treaty controlling marine pollution in the north-east Atlantic Ocean around Europe. A similar Barcelona Convention exists to protect the Mediterranean Sea. The Water Framework Directive of 2000 is a European Union directive committing EU member states to make their inland and coastal waters free from human influence.

In the United Kingdom, the proposed Marine Bill is designed to "ensure clean healthy, safe, productive and biologically diverse oceans and seas, by putting in place better systems for delivering sustainable development of marine and coastal environment".

Under the umbrella of UNEP, numerous cooperative efforts have been held to reach protocols and conventions. For instance, a Protocol on Integrated Coastal Zone Management was approved in January 2008, involving 21 countries bordering the Mediterranean Sea, as well as the European Union. Within the framework of Land Based Sources Protocol for pollution reduction from land-based sources, Mediterranean countries and parties to the Barcelona Convention have agreed this year on an initial set of actions covering the reduction of municipal pollution and the elimination of a number of Persistent Organic Pollutants.

The Caribbean Environment Programme (CEP) continues to encourage member states in meeting the Caribbean Challenge target of protecting 20 percent of marine and coastal habitats by 2020. The Caribbean Large Marine Ecosystem Project and development of a Regional Fund for Wastewater Management will support regional collaboration to reduce the vulnerability of sensitive coastal and marine ecosystems by improving national and regional governance structures and developing new and innovative mechanisms for financing new pollution reduction activities.

Even though the greatest problem with international legislation is its actual enforcement, the efforts toward concerted actions can only be promoted.

Nationally

A strict Chinese limit on ultra-thin plastic bags significantly reduced bag-related pollution nationwide during the past year. "Our country consumes a huge amount of plastic shopping bags each year" a spokesperson for China's State Council said, when announcing the ban last May. "While plastic shopping bags provide convenience to consumers, this has caused a serious waste of energy and resources and environmental pollution because of excessive usage, inadequate recycling and other reasons." In January 2008, The State Council, China's parliament, passed legislation to prohibit shops and supermarkets from providing free plastic bags that are less than 0.025 millimeters thick. The State Administration of Industry and Commerce also threatened to fine shopkeepers and vendors as much as 10,000 Yuan ($1,465) if they were caught distributing free bags. The country avoided the use of 40 billion bags, according to government estimates. The National Development and Reform Commission (NDRC) estimated that the limit in bag production saved China 1.6 million tons of petroleum.

The first country to ban plastic bags was Bangladesh, which did so in 2002. Following a particularly damaging typhoon, authorities discovered that millions of bags were clogging the country's system of flood drains, contributing to the destruction.

In the same year, Ireland took another approach and instituted a steep tax on plastics. According to the country's Ministry of Environment, use fell by 90 percent as a result and the tax money that was generated funded a greatly expanded recycling program throughout the country. In 2003, the government of Taiwan put in place a system by which bags were no longer made available in markets without charge. Carryout restaurants were even required to charge for plastic utensils.

Larger economies have joined the cause and passed legislations on a national level. In 2005, French legislators imposed a ban on all non-biodegradable plastic bags, which will go into effect in 2010. Italy will also ban them that year.

During its 2008 session, the New York State Legislature passed legislation requiring the reduction, reuse, and recycling of checkout bags. The previous year, the city of San Francisco banned plastic bags altogether, at least the flimsy ones of yore. National Public Radio reported a few months later that the ban had been a boom for local plastics manufacturers, who have been introducing heavy-duty, recyclable, and even compostable bags into the marketplace.

Media and Creative Awareness

An impactful vehicle for information and awareness is indubitably found in the media and creative ventures.

A good example of such ventures is the team of two South African surfers, Ryan and Bryson Robertson, and one Canadian, Hugh Patterson, who created the OceanGybe mission. Their plan is to circumnavigate the globe in a small 40ft sailboat and surf remote reef breaks on far flung islands while interacting with the local cultures. They intend to spread awareness of the vast tracts of plastic and trash afloat on the world's oceans that inevitably ends up on some unsuspecting shore.

More publicized and funded is the environmentalist and Adventure Ecology founder David de Rothschild's expedition: the Plastiki mission.

The Plastiki, a one-of-a-kind 60-foot catamaran, was created out of 10,000 reclaimed plastic soda bottles, self-reinforced PET (polyethylene terephthalate) and recycled materials. The vessel's name is a nod to famed explorer Thor Heyerdahl, who led a 1947 voyage on the Kon-Tiki to test theories of Polynesian settlement by South Americans. The Plastiki is about to make its momentous voyage across the Pacific Ocean, a 10,000-mile expedition from San Francisco to Sydney, Australia by the end of this year, to inspire people to rethink current uses and waste of plastic as a resource and bring attention to the GGP.

De Rothschild explained that Plastiki's construction has already jump-started research into a future "smart plastics" industry before ever leaving port. For instance, studies are underway on glues that could someday replace common marine epoxies and plastics that could replace non-recyclable fiberglass.

"The Plastiki voyage will be a great adventure, but I think more exciting is the ability to create a conversation on the issue of plastics."

Philosophy

Adventures of philosophical nature have been taking place as well.

Indeed, French thinkers such as Michel Serre or Luc Ferry, The new ecological order, have developed a train of thought aiming towards a legal recognition, therefore legal protection of Nature. This type of philosophy has been called, deep ecology. The principle is quite simple: democracies have installed their legislative framework, their "social contract," omitting Nature as a protagonist/subject of law. Therefore, to protect Nature, i.e. our environment, should we confer legal right to it, thus making nature a legal subject/person?

Obviously, all subject of law have rights, but they also have obligations. If we can easily forsee what the right and protection would be for this legal subject, what would be its obligations?

This leads many thinkers towards a notion of "droit ou devoir d'ingerence ecologique" (right or duty of intervention/assistance), trying to mirror the situation on the humanitarian field. The notions of "self-defense" and "non-assistance a personne en danger" have also been explored as possible legal frames to better enforcement of laws and conventions aimed to protect the environment, and curb ocean plastic pollution for that matter.

Sustainable and Future Technologies – Opportunities and Innovations

Biodegradable Plastics

Biodegradable plastics have been considered as a future, sustainable option to curb our voracious demand and consumption of plastic material as known in its current form. According to the Biodegradable Plastics Society, when such plastics are composted they break down to carbon dioxide and water.

Controversy does exist though, because it is possible that biodegradable plastics do not break down fully, especially under environmental conditions which are not ideal for composting, and leave non-degradable constituents, some of which may be equally, if not more, hazardous. Also, there is a danger that biodegradable plastics will be seen as "litter friendly" materials, conveying the wrong message to the public and potentially leading to less responsible and more wasteful practices.

A change in behavioral propensities to over-consume plastics, discard and thus pollute, need to be promoted to the fullest.

Ongoing Discoveries and Solutions to the Traditional Plastic Waste Problem

Scientists have been searching for solutions to the traditional plastic waste problem.

In 2008 and 2009, two high school students who discovered plastic-consuming microorganisms, might have found groundbreaking solutions.

The first was Daniel Burd. The second was Tseng I-Ching, a high school student in Taiwan.

Daniel's simple and clever process was to immerse ground plastic in a yeast solution that encourages microbial growth, then isolating the most productive organisms. After several weeks of tweaking and optimizing temperatures, Burd was achieved a 43 percent degradation of plastic in six weeks, an almost inconceivable accomplishment. It appeared as an environmentalist's dream: a non-chemical, i.e. fully organic, low cost and nontoxic method for degrading plastic.

There have been several successful bacteria based solutions developed at the Dept. of Biotechnology in Tottori, Japan, as well as at the Dept. of Microbiology at the National University of Ireland, but both apply only to styrene compounds.

Similarly, a 2004 study at the University of Wisconsin isolated a fungus capable of biodegrading phenol-formaldehyde polymers previously thought to be non-biodegradable.

Green Chemistry and "Begnign by Design" Concept

A growing interest amongst chemists, and ultimately industries, is Green chemistry- policy, also called "benign by design".

According to scientists at the University of Southern Mississippi (USM), a new type of environmentally friendly plastic that degrades in seawater may be developed. Robson F. Storey, Ph.D., a professor of Polymer Science and Engineering at USM, said, "We're moving toward making plastics more sustainable, especially those that are used at sea." Their study is funded by the Naval Sea Systems Command (NAVSEA), which is supporting a number of ongoing research projects aimed at reducing the environmental impact of marine waste. The new plastics are made of polyurethane that has been modified by the incorporation of PLGA [poly (lactide-co-glycolide)], a known degradable polymer used in surgical sutures and controlled drug-delivery applications. When exposed to seawater, the plastics degrade via hydrolysis into nontoxic products, according to the scientists. The plastics are not quite ready for commercialization. "More studies are needed to optimize the plastics for various environmental conditions they might encounter, including changes in temperature, humidity and seawater composition", Storey says.

A new kind of material, called oxo-biodegradable plastic, does not just fragment, but is consumed by microorganisms after the additive has reduced the molecular weight. It is thus biodegradable. This process continues until the material has biodegraded to nothing more than CO_2, water, humus, and trace elements. There is little or no additional cost, as it can be made with the same machinery and workforce as conventional plastic. The time taken to degrade can be programmed to a few months or a few years and, until the plastic degrades, it has the same strength and other characteristics as conventional plastic. Oxo-biodegradable plastic will be engineered to degrade in a short time leaving no harmful residues.

Recycling and Zero Waste Concept

A promising way toward a future of better plastic waste management is recycling the material. The recycling industry might eventually be a path leading to considerable opportunities and solutions.

The BIR (Bureau of International Recycling), whose headquarters is in Belgium, is a trade federation representing the world's recycling industry. About 800 companies and national federations from over 70 countries are affiliated with the BIR. Together they provide their expertise to other industrial sectors and political groups in order to promote recycling. It is estimated that the recycling industry employs more than 1.5 million people, annually processes over 500 million tons of commodities, and has a turnover exceeding $160 billion.

However, this industry is faced with many challenges, as the recycling material itself is very diverse in a chemical sense and can release, when processed, extremely dangerous chemicals. For instance, a recycling factory in China was recently exposed to tragic consequences due to the recycling of very hazardous plastic materials. It was reported that a team of workers in China's Zhejiang province collapsed after handling two metric tons of plastic scrap on September 13, 2009. At least

21 have since been hospitalized and three of them have died. According to the initial investigative conclusions, the victims were in contact with highly toxic chemical, dinitrophenol, which was found on the two tons of plastic scrap. Workers at the recycling factory were unaware of the hazard of the material and had no protection during the unloading. This particular tragedy is only the tip of the iceberg. China's plastics recycling industry is poorly regulated, with scandals such as biohazard plastic waste being melted and reprocessed into consumer goods.

Recycling is definitely a potentially great path to solving the plastic waste problem but definitely not the most unchallenging one.

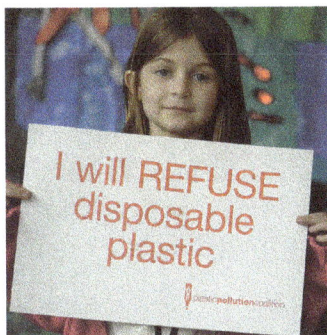

Along the same lines, a responsible waste strategy, namely the concept of Zero Waste, has been widespread. Such a strategy encompasses waste reduction, reuse and recycling as well as producer responsibility and Eco design. According to a Greenpeace report, strategies to achieve Zero Waste are adopted throughout the world, in industrialized countries and in less developed countries.

Ultimately, this would mean reduction of the use of plastics. "Our understanding of disposal and reuse (of plastic, is what) is to blame." as many environmentalist such as de Rothschild, said.

This zero waste philosophy encourages the redesign of resource's life cycles, so that all products are reused. Any trash sent to landfills is minimal. The process recommended is one similar to the way that resources are reused in nature. Zero waste can represent an economical alternative to waste systems, where new resources are continually required to replenish wasted raw materials.

DTSC's Environmental Chemistry Laboratory is currently analyzing some of the plastic marine debris collected at the Great Garbage Patch by Project Kaisei scientists, and explores the potential of converting the plastic collected into new material.

Indeed, Doug Woodring from Project Kaisei stated last September that they intend to use some of the newest plastic technologies to detoxify and turn the plastic waste caught in the oceans either into fuel or another useable material. Thus, Project Kaisei hopes to assign value to that plastic collected, particularly the overwhelming majority that is never recycled. It becomes obvious that technologies that convert plastic to fuel, clothing, or simply more profitable plastic could give people a good reason to pick up all that plastic and make a profit from it. Numerous industries, such as fashion, are already increasingly focusing on new green materials as a base for their offered products, encouraging a way of life and cultural change toward better choices and awareness of the environment.

Methods to Control Pollution

There are Various Methods to control pollution like; devices to prevent air pollution, Soil pollution can also be controlled by various means and methods such as landfill, composting etc. and Water pollution is also reduced and controlled in the same way in agriculture by following certain measures and methods that are eco-friendly. Some of the effective and practical control measures for minimizing environmental pollution are outlined below:

1. Combustible solid wastes should be burnt in incinerators. This method does not solve the problem in a real sense because in this, solid waste is being converted into gaseous wastes causing air pollution. Unless it is properly controlled, incineration may cause more nuisances.

2. Solid organic wastes including faecal matter and wastes from tanneries should be converted into compost manure at the places far away from the cities and human dwellings. The composting should be done in pits or in heaps adequately covered with layers of soil at least 8-10 cm thick to prevent fly breeding and rat menace which are important carriers of various diseases.

3. Non-combustible solid waste materials like ash, rubbish, tins, glass pieces if not recoverable for usual purposes should be disposed of by landfill method in low-lying areas.

4. Anaerobic septic tank treatment can be used for individual houses or small communities. Besides, aerobic biological treatment systems including trickling filters, activated sludge treatment and oxidation ponds can also be used for liquid wastes or sewage disposal.

5. Automobiles must be either made to eliminate use of gasoline and diesel oil or complete combustion is obtained in the engine so that noxious compounds are not emitted. The automobiles, trucks and other transport systems must have an antismog device. In some countries factories are using devices like scrubbers, cyclone separators or electrostatic precipitators to minimize pollution.

6. There should be cut back in the use of fertilizers, herbicides, pesticides and other agrochemicals as far as possible.

7. Excessive and undesirable burning of vegetation should be stopped.

8. Sponges and towels should be used in place of paper towels and also the use of paper cups and plates and similar materials should be stopped.

9. Little use of electric appliances and motor-nm appliances will reduce thermal pollution.

10. Washing soda and scouring pad should be used instead of detergents.

11. Waste management is based on principle of '3Rs' i.e. Reduce, Recycle and Reuse. Used boxes, bags, plastics and bottles should be reused whenever possible.

12. Since about 40% of the phosphates in water pollution come from detergent, it has been suggested that only detergents low in phosphates should be used.

13. Shampoos, lotions and similar products should not be bought in plastic bottles. It has recently been suggested that use of plastic containers and glasses may cause cancer.

14. Smoking should be stopped (there is 5, 00,000 tonnes tobacco pollution annually).

15. Proper attention should be given by the government to make people realize the implications of environmental problem.

16. Legislation against pollution should be strictly implemented.

17. International action is needed to deal with the problems presented by highly toxic pollutants like lead, mercury, organ chlorine pesticides released in to the atmosphere and carried far beyond the country of origin as well as carried down to the sea by rivers. Successful action to improve environmental qualities depends mainly on the acceptance by industry and local authorities of the need to reduce greatly both quantities and toxicity of certain wastes at present being discharged into the sea and the modernization and expansion of sewage disposal systems.

18. Environmental education is the best programme to deal with the environmental problems. It is most fundamental in our efforts to combat and control pollution, over-population and misuse of natural resources.

Reduce

The first and most effective component of the waste hierarchy is reducing the waste created. Consumers are encouraged to reduce their waste by purchasing in bulk, buying items with less packaging and switching to reusable instead of single-use items. Businesses can adopt manufacturing methods that require fewer resources and generate less waste. In addition to benefiting the environment, these efforts often offer consumers and businesses the financial incentive of lower expenses in purchases.

Tips for Reducing Waste

At Home

- Buy items in bulk or as refills in order to reduce packaging waste. Plan what you will need before you shop to ensure you buy only what you will use, so that no items are wasted. Purchase items that are durable and long-lasting, so they can be utilized over and over.

- Clean using washable rags and eat using cloth napkins to reduce the waste of paper towels and napkins.

- When it comes time for spring-cleaning, donate unwanted or outgrown items to local charities, churches or community centers. Or, check Free cycle, an online network of which Waste Management is a partner where members give away and get items for free in their own communities - rather than filling up our landfills with gently used items, find someone else who could put it to good use.

- Look for opportunities to operate your home more efficiently - for example, turn off lights when you are not using them to extend the life of light bulbs, so used ones are thrown out less often.

- Cut down on junk mail by contacting the companies or catalogues that send you the pieces and asking them to remove your home from their list. Visit the Direct Marketing Association's Web site to request that your name be removed from mailing lists. If you receive catalogues you no longer read or repeated solicitations from the same company, contact them directly to request removal from their marketing lists - and instead, check out their goods online.

- Pay your bills online - it saves the paper utilized in generating a hard copy bill, as well as stamps.

- Donate recent and gently used magazines or books that you no longer want to local medical facilities, retirement communities or schools. Contact the facilities to see what they might be able to use.

- Pack kids' lunches in re-useable containers that will keep items cool and also cut down on trash. If items within their lunch boxes are recyclable, such as juice containers, water bottles or yogurt cups, make sure they go in the recycle bins at school - or if their school does not have a recycling program, bring them home.

- Ask your grocery store if they have a plastic bag recycling program. Or better yet, use a sturdy cloth shopping bag or request paper bags at the store, which can be recycled along with your newspapers and other paper goods.

- Request that your dry cleaner not wraps your clothing in plastic, or see if he or she will accept the bags back to recycle. Take any extra hangers you receive from the cleaners that would otherwise be thrown out back to them - see if the dry cleaner can reuse them.

- Wrap birthday and holiday gifts with recyclable paper or re-use materials from around the house, such as the comics pages from the newspaper or paper from a previous gift. Minimize the use of bows or other decorative items, which cannot always be re-used and instead end up in the trash.

- Use similar materials - brown paper shopping bags, department store bags, old posters or sturdy wrapping paper - to cover your children's schoolbooks.

- Create fun craft projects for your children utilizing items that might otherwise be thrown out - for example, use broken or mismatched buttons to make mosaics or picture frames, turn scraps of fabric, yarn or beads into jewelry or hair decorations and decorate empty, clean containers like oatmeal tubs, lunch meat containers, pasta sauce jars or cookie tubs to store crayons, toys or collectibles.

- When packing items for shipping, minimize the use of polystyrene foam "popcorn," which cannot be recycled and often creates litter when thrown away. Utilize a padded mailer or newspaper, both of which can be used multiple times. If you receive items with popcorn, contact your local shipping center to see if they will re-use it - or, if it needs to be thrown away, put it in a plastic bag and tie the ends so it will not blow away.

- Utilize re-chargeable batteries around the house. Remember, batteries cannot be thrown into the trash, so be sure to properly dispose of them at a household hazardous waste center.

At Work

- Utilize recycled paper for printing and copying, and recycle the paper you would otherwise throw out.

- If you need to print out large documents, print on both sides.

- Buy refillable toner and printer cartridges.

- Use refillable pens and pencils.

- Support companies that sell recyclable office goods or minimize the packaging of their products.

- When possible, correspond by phone or e-mail and keep your files electronic, rather than as hard copies. If you have an electronic organizer or PDA, store contact information or important messages here rather than printing it out.

- When faxing, use a stick-on address label instead of a cover sheet and program your fax machine to eliminate the production of confirmation sheets.

- If possible, utilize software to route faxes to your computer, rather than as printouts.

- If you subscribe to magazines or newspapers at the office, utilize a routing slip so that multiple people can read the same copy of each publication. Or, consider buying an online subscription to the publication in order to read the content on the Web.

- Bring leftovers from home for lunch, in re-useable containers. Avoid the excess packaging and waste created by fast-food or take-out, and eat healthier, too.

- Get your coffee or water in a mug or travel container, rather than a paper or Polystyrene foam cup. If you drink water from a plastic water bottle, refill it and continue to use it in order to extend its life. Be sure to recycle it when you are done.

- If your office does not have a recycling program for items such as cans, plastic bottles, glass containers or paper, speak to your colleagues about starting one. Contact the solid waste company who services your office to discuss recycling options.

- Install regulated or motion-activated towel dispensers in the restrooms so that paper towels are not wasted.

- If your office is going through a remodel or has surplus furniture or supplies, look for opportunities to donate re-useable items to local charities, community groups or to employees for use at home. Work with the construction company and your solid waste provider to recycle or re-use the materials from the re-model.

Reuse

Reuse, is that you should reuse items as much as possible before replacing them. For example, it generally makes more environmental sense to update your computer rather than get rid of it and buy a new one. However, if you do replace your computer, you should ensure that it, or its components, is reused. Many charitable organizations welcome donations of second-hand computers.

Reuse is a means to prevent solid waste from entering the landfill, improve our communities, and increase the material, educational and occupational wellbeing of our citizens by taking useful products discarded by those who no longer want them and providing them to those who do. In many cases, reuse supports local community and social programs while providing donating businesses with tax benefits and reduced disposal fees.

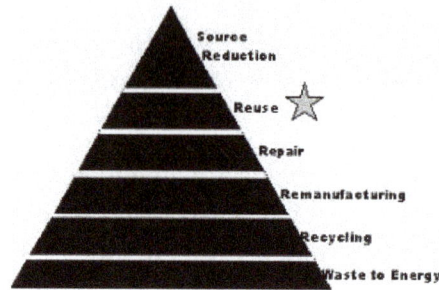

Environmental Benefits

Many reuse programs have evolved from local solid waste reduction goals because reuse requires fewer resources, less energy, and less labor, compared to recycling, disposal, or the manufacture of new products from virgin materials. Reuse provides an excellent, environmentally-preferred alternative to other waste management methods, because it reduces air, water and land pollution, limits the need for new natural resources, such as timber, petroleum, fibers and other materials. The US Environmental Protection Agency has recently identified waste reduction as an important method of reducing greenhouse gas emissions, a contributing factor to global warming.

Community Benefits

For many years, reuse has been used as a critical way of getting needed materials to the many disadvantaged populations that exist. Reuse continues to provide an excellent way in which to get people the food, clothing, building materials, business equipment, medical supplies and other items that they desperately need. There are other ways, however, that reuse benefits the community. Many reuse centers are engaged in job-training programs, programs for the handicapped or at-risk youth programs.

Economic Benefits

When reusing materials, instead of creating new products from virgin materials, there is less burden on the economy. Reuse is an economical way for people of all socio-economic circles to acquire the items they need. From business furniture to household items, from cars to appliances, and just about anything else you could think of -- it is less expensive to buy used than new.

Reuse Supports Solid Waste Management Goals

Buying and using items that are reusable supports a method of waste management. In many cases an item can be reused several times, and then sent to the recycling center for processing. The list of reused items is virtually unlimited, and reuse centers can be found in nearly every community.

- Building Materials: Lumber, tools, windows, doors, light fixtures, paint, plumbing supplies and fixtures, architectural pieces, fencing, hardware, and many other items needed for constructing or refurbishing a building can be found used.

- Office Furniture and Supplies: Desks, tables, chairs, filing cabinets, credenzas, shelving units, stacking trays, tape dispensers, notebook binders and other equipment and supplies can be reused in offices, schools, hospitals, non-profit organizations and others.

- Computers and Electronics: Personal computers, printers, fax machines, televisions, video cassette recorders can be reused in business, personal, and non-profit environments.

- Art Materials: Fabric, paint, lumber, stage props, and a wide variety of other items can be used for school or cultural organization creative projects.

- Medical Equipment and Supplies: Equipment and supplies that are obsolete to one hospital, clinic or organization may find a home in another facility, especially those in less-industrialized nations.

- Surplus Food Items and Equipment: Boxed, bagged, canned and even prepared food from grocery stores, warehouses, manufacturers' over-runs and discontinued items, catered events, restaurants can be reuse by homeless shelters, soup kitchens, and other organizations serving disadvantaged people. Stoves, refrigerators, freezers, and other items can also be found used.

- Household Items: Appliances, clothing, furniture, dishes, vehicles, paint, and virtually anything else for the home can be found by shopping reused instead of brand new. And in most cases, at a significantly lower price. Another form of reuse is shopping specialty stores that sell antiques or vintage items. Shopping and holding garage and yard sales are other popular forms of reuse.

Recycle

Recycling is processing used materials (waste) into new, useful products. This is done to reduce the use of raw materials that would have been used. Recycling also uses less energy and great way of controlling air, water and land pollution.

Effective recycling starts with household (or the place where the waste was created). In many serious countries, the authorities help households with bin bags with labels on them. Households then sort out the waste themselves and place them in the right bags for collection. This makes the work less difficult.

Waste items that are usually recycled include:

Paper Waste

Paper waste items include books, newspapers, magazines, cardboard boxes and envelopes.

Plastic Waste

Items include plastic bags, water bottles, rubber bags and plastic wrappers.

Glass Waste

All glass products like broken bottles, beer and wine bottles can be recycled.

Aluminium Waste

Cans from soda drink, tomato, fruit cans and all other cans can be recycled.

When these are collected, they are sent to the recycling unit, where all the waste from each type are combined, crushed, melted and processed into new materials.

Importance and Benefits of Waste Recycling

Recycling Helps Protect the Environment

This is because the recyclable waste materials would have been burned or ended up in the landfill. Pollution of the air, land, water and soil is reduced.

Recycling Conserves Natural Resources

Recycling more waste means that we do not depend too much on raw (natural) resources, which are already massively depleted.

Recycling Saves Energy

It takes more energy to produce items with raw materials than from recycling used materials. This means we are more energy efficient and the prices of products can come down.

Recycling Creates Jobs

People are employed to collect, sort and work in recycling companies. Others also get jobs with businesses that work with these recycling units. There can be a ripple of jobs in the municipality.

References

- Pollutiontypes: conserve-energy-future.com, Retrieved 11 April 2018

- Types-of-environmental-pollutants-221368: livestrong.com, Retrieved 10 May 2018

- Causes-effects-solutions-of-air-pollution: conserve-energy-future.com, Retrieved 11 July 2018

- What-is-noise-pollution, noise-pollution: eschooltoday.com, Retrieved 28 May 2018

- Agricultural-pollution-causes-effects-types-prevention-methods: naturalenergyhub.com, Retrieved 12 March 2018

- Meaning-three-rs-reduce-reuse-recycle-79718: homeguides.sfgate.com, Retrieved 28 June 2018

- What-is-recycling, waste-recycling: eschooltoday.com, Retrieved 09 March 2018

8

Environmental Management

The control and management of the human influence on the environment is under the scope of environmental resource management. It strives to ensure the protection and maintenance of ecosystem services for use by human societies in the future, while ensuring ecosystem integrity. This chapter discusses in extensive detail the fundamentals of water resource management, solid waste management, air resource management and environmental management.

Impact of Pollution on Environment

The pollution of the environment & its natural resources such as water, air or land with different pollutants is known as environmental pollution. The biggest & main harmful effect of pollution is on the environment as it breaks up the environment & also the different ecosystems present in it. Environmental pollution has adverse effects on both the humans & the other environmental living and non-living things. Environmental pollution is a worldwide problem & it causes hazardous effects on humans & natural resources. Environmental pollution is defined as the state of contamination of different natural resources of the environment with the introduction of the poisonous chemicals & gases in the atmosphere of the earth which leads towards the destruction of natural resources of the environment such as land, air or water. The different pollutants which pollute the environment may be regarded as primary or secondary pollutants & the pollutants having short term or long term effects on the environment due to their vitality & nature of causing damage to the environment. It is the state of the buildup of toxic chemicals & poisonous gases in the breathing zone of the atmosphere of the earth which leads to many harmful disorders & discomforts to all the life species relying on natural resources of the environment. Environment pollution occurs in pollution of different forms of the environment such as land, water, air, noise, thermal, radioactive or light pollution. When the pollutants enter in the different zones of the environment, the species dependent on these natural resources would suffer & face difficulties in surviving. The environment is polluted when the different types of pollutants such as greenhouse gases, harmful heavy metals & harmful chemicals. The pollutants cause the long term as well short-term changes in the environment which have very dangerous effects.

Environmental Impact of the Coal Industry

Mining is the first step in the dirty life cycle of coal. When coal mines move in, whole communities are forced off their land by expanding mines, coal fires, subsidence, and overused and contaminated water supplies. Mines are quick to dig up and destroy forests and soils. But once the coal is gone, the problems they leave behind, like acid mine drainage, can persist for decades. Around the world, Greenpeace campaigns to help communities stop coal mines, and speed up the shift to 100 percent clean, safe renewable energy.

Underground mines, which provide the majority of the world's coal, allow coal companies to extract deep coal deposits. About 40 percent of the world's coal mines are the more damaging strip mines (also called open cast, open pit, mountaintop or surface mining).

Strip Mining Impacts

Strip mining is highly destructive. Yet the industry often prefers to strip mine because it takes less labor and yields more coal than underground mining. In some countries, such as Australia, strip mines make up 80 percent of mines.

Strip Mining Damages and Pollutes Ecosystems

Strip mining clears trees, plants and topsoil. Mining companies scrape away earth and rocks to get to coal buried near the surface. Mountains may be blasted apart to reach thin coal seams within, leaving permanent scars on the landscape.

In this way, strip mining destroys landscapes, forests and wildlife habitats. It leads to soil erosion and destruction of agricultural land.

When rain washes topsoil disturbed by mining into streams, these sediments pollute waterways. This can hurt fish and smother plant life downstream. It can also disfigure river channels and streams, which leads to flooding.

Strip mining also causes noise pollution and dust as heavy machinery disrupts topsoil and mining activity creates coal dust.

Strip Mining Contaminates Water

When miners upturn earth, minerals and heavy metals within it can dissolve into mine wastewater and seep into the water table. This increases risk of chemical contamination of groundwater and acid mine drainage.

Strip mining also lowers groundwater levels around the mine. This is because, in order to remove coal, vast quantities of groundwater must be pumped out of the mine. As a result, surrounding ecosystems and farmland may become drier, and erosion may start to change the landscape. Strip mining also uses significant amount of water to suppress dust.

When mines lower groundwater levels, this also affects local people, who must continually drill deeper wells to get water.

Washing coal (to remove unwanted materials) creates toxic waste slurry that can threaten surface waters or leak into groundwater.

Coal power plants also strain precious global water supplies.

Strip Mines Leave Lands Barren

Coal mining is land disturbance on a vast scale.

- In the US, from 1930 to 2000, coal mining altered about 2.4 million hectares (5.9 million acres) of natural landscape, most originally forest.

This mining activity leaves behind barren lands that stay contaminated long after the mine shuts. Although many countries require coal mines to have reclamation plans, it is a long, difficult task to undo all their damage to water supplies, habitats and air quality. Re-seeding plants is difficult because mining thoroughly damages soil. If coal companies go bankrupt, costly rehabilitation may be left undone.

- In China, coal mining degraded the quality of 3.2 million hectares of land, according to a 2004 estimate, but total mine wasteland was restored at a rate of only 10 to 12 percent.

- In Montana, US, replanting projects were only 20 to 30 percent successful. In Colorado, even lower survival (about 10 percent in some locations) was seen for oak aspen seedlings.

Underground Coal Mining Impacts

Although seen as less destructive than strip mining, underground mining still causes widespread damage to the environment.

Subsidence

Collapse of earth into underground mines, or subsidence, is a serious problem.

In room-and-pillar and long-wall mines, columns of coal and other structures are used to support the ground above. Later in the mining process, they are often taken out. The mines are left to collapse. The land above starts to sink, seriously damaging buildings and entire landscapes. Subsidence can also cause farmland to fill with water and become wetland or lakes.

Underground Mine Water Drained Away

Underground mining lowers the water table, changing the flow of groundwater and streams.

In Germany, the mining industry pumps over 500 million cubic meters of water out of the ground every year. Only a small percentage of this water is used by industry or local towns — the rest is wasted. What's worse, removing so much water creates a kind of funnel that drains groundwater from an area that is much larger than the immediate coal-mining environment.

Underground Mines Bring Toxins to Surface

Underground mining also brings huge amounts of waste earth and rock to the surface. This waste often becomes toxic when it contacts air and water.

Coal Mine Methane

Coal mining releases methane into the atmosphere. Formed during the geological process that creates coal, methane is 84 times as powerful as carbon dioxide at disrupting the climate over a 20-year timespan.

Globally, about six percent of methane emissions due to human activity come from coal mining.

Most coal mine methane comes from underground mines. This methane is often captured and used as town fuel, industrial fuel, chemical feedstock and vehicle fuel. Methane is also used in power generation projects.

The process to extract this methane, coal seam gas fracking, creates large amounts of waste water, risking surface and groundwater sources. It also increases the risk of uncontrolled methane leaks, contaminating water sources and destroying climate. Yet coal bed methane projects have been increasing rapidly globally.

Coal Fires Smoulder and Pollute

Coal fires can burn for decades or even centuries, releasing fly ash and smoke laden with greenhouse gases and toxic chemicals. These fires are a significant environmental problem in China, Russia, the US, Indonesia, Australia and South Africa.

Coal fires occur when coal seams burn or smoulder, or when coal storage or waste piles burn. Lightning, forest fires and peat fires can start coal fires. But they are often caused by mining accidents and bad mining practices. In Indonesia, the same fires used to clear large tracts of rainforest ignited over 300 coal fires since the 1980s.

Underground coal fires can release smoke laden gases including carbon monoxide (CO), carbon dioxide (CO_2), methane (CH_4), and sulphur dioxide (SO_2). Coal fires also cause fly ash to release from mine vents and fissures.

Coal fires can cause temperatures to rise at the surface, and contaminate groundwater, soil and air.

China has the world's most coal fires. Between 20 and 200 million tons of coal burn uncontrollably each year. This accounts for 0.5 to 5 percent of China's national coal consumption and related carbon dioxide emissions. (Although coal fires are significant, emissions from China's power plants are far higher.) India, on the other hand, has the world's greatest concentration of coal fires.

Acid Mine Drainage

When coal and other rocks unearthed during mining mix with water, this creates acid mine drainage. The water takes on toxic levels of minerals and heavy metal and leaks out of abandoned mines. From there it contaminates groundwater, streams, soil, plants, animals and humans.

Taking on an orange color, it can blanket rivers, estuaries or sea beds, killing plants and making surface water unusable for drinking. Acid mine drainage can continue for decades or centuries after a mine closes unless costly reclamation projects are done.

Greenpeace documented massive open-cast coal mines' harmful effects in Kalimantan, Borneo. The mines cause widespread water pollution when they discharge toxic waste into rivers and leave acid mine drainage to collect in artificial lakes.

Coal Mining Harms Workers' and Residents' Health

Mining coal, the dirtiest fossil fuel on the planet, exposes both miners and local populations to health hazards.

Threat to Mine Workers

When people who work in mines, or live close by them, inhale coal dust and carbon, this hardens their lungs, leading to black lung disease (also called pneumoconiosis or CWP). An estimated 1,200 people in the US still die from black lung disease annually. The situation is even worse in developing countries.

Mine collapses and accidents kill over a thousand workers around the world every year. Chinese coal mine accidents killed more than 900 people in 2014 alone.

Threats to Local Populations

People living near coal mines have higher-than-normal rates of cardiopulmonary disease, chronic obstructive pulmonary disease, hypertension, lung disease, and kidney disease.

Local communities also suffer when coal fires occur. These fires emit toxic levels of arsenic, fluorine, mercury and selenium, contaminants that can enter the air and food chain of local communities.

Environmental Impact of Nuclear Power

Nuclear energy has been proposed as an answer to the need for a clean energy source as opposed to CO2-producing plants. Nuclear energy is not necessarily a clean energy source. The effects nuclear energy have on the environment pose serious concerns that need to be considered, especially before the decision to build additional nuclear power plants is made.

Carbon Dioxide

Nuclear power has been called a clean source of energy because the power plants do not release carbon dioxide. While this is true, it is deceiving. Nuclear power plants may not emit carbon dioxide during operation, but high amounts of carbon dioxide are emitted in activities related to building and running the plants. Nuclear power plants use uranium as fuel. The process of mining uranium releases high amounts of carbon dioxide into the environment. Carbon dioxide is also released into the environment when new nuclear power plants are built. Finally, the transport of radioactive waste also causes carbon dioxide emissions.

Low Level Radiation

Nuclear power plants constantly emit low levels of radiation into the environment. There is a differing of opinion among scientists over the effects caused by constant low levels of radiation. Various scientific studies have shown an increased rate of cancer among people who live near nuclear power plants. Long-term exposure to low level radiation has been shown to damage DNA. The degree of damage low levels of radiation cause to wildlife, plants and the ozone layer is not fully understood. More research is being done to determine the magnitude of effects caused by low levels of radiation in the environment.

Radioactive Waste

Radioactive waste is a huge concern. Waste from nuclear power plants can remain active for hundreds of thousands of years. Currently, much of the radioactive waste from nuclear power plants has been stored at the power plant. Due to space constraints, eventually the radioactive waste will need to be relocated. Plans have been proposed to bury the radioactive waste contained in casks in the Yucca Mountains in Nevada.

There are several issues with burying the radioactive waste. Waste would be transported in large trucks. In the event of an accident, the radioactive waste could possibly leak. Another issue is uncertainty about whether the casks will leak after the waste is buried. The current amount of radioactive waste requiring long-term storage would fill the Yucca Mountains and new sites would need to be found to bury future radioactive waste. There is no current solution to deal with the issue of radioactive waste. Some scientists feel that the idea of building more nuclear power plants and worrying about dealing with the waste later has the potential of a dangerous outcome.

Cooling Water System

Cooling systems are used to keep nuclear power plants from overheating. There are two main environmental problems associated with nuclear power plant cooling systems. First, the cooling system pulls water from an ocean or river source. Fish are inadvertently captured in the cooling system intake and killed. Second, after the water is used to cool the power plant, it is returned to the ocean or river. The water that is returned is approximately 25 degrees warmer than the water was originally. The warmer water kills some species of fish and plant life.

There is no disagreement that clean sources of energy are vital to the environment. The disagreement lies in what form that clean energy should be in. Supporters of nuclear energy argue that it is an efficient source of energy that is easy to implement. People against nuclear energy propose using combined methods of solar, wind and geothermal energy. Solar, wind and geothermal energy still have environmental issues, but ones that are not as great as nuclear plants or coal-burning power plants.

Environmental Impact of Cleaning Agents

Only small percentages (7%) of cleaning products disclose enough information to realistically estimate both their carbon footprints and environmental impact. In an increasingly ecologically minded society, this is surprising. With so little information, this means that consumers cannot make an informed decision, authorities cannot accurately track the health and environmental impact, and activists cannot pressure companies to produce a more eco-friendly product.

Environmental Impact of Cleaning Products

Only a few of us actually look at product labels to assess the carbon footprint and environmental impact. Current estimates suggest the carbon footprint producing these cleaning products to be 0.7lbs of CO_2 per pound of product. In terms of volume of CO_2, this is roughly the size of a 2 foot cube filled with CO_2.

This is a stark contrast to the food industry in Britain, where every action and emission is assessed. With regard to cleaning products, we do not know the cost of processing the chemicals, packaging them, creating the packaging, or transportation.

Carbon footprint only covers the amount of energy required to produce the product. However, its impact goes well beyond this to what leaked, spilled or used products have on the environment. For example, APEs, a type of cleaning compound with bioactive consequences has been banned in the UK for its effect on an organism's endocrine system. Cleaning wipes are clogging up drains, rivers, oceans, and seafood across the world. Furthermore, many cleaning products contain toxic substances such as formaldehyde, ammonia, and sodium lauryl sulfate.

The Health Impact of Cleaning Products

It is always good to know exactly what chemicals are in your cleaning products because they can impact many adults, and especially children and pets. Fortunately, in many cases, the sheer size of an adult makes them less affected by these products but also our habits do too. Animals and children are more tactile, meaning they interact with the world through touch and taste as well as using eyes, noses, and mouths. Let's look at some common ingredients in many cleaning products.

- Formaldehyde: High doses of formaldehyde can irritate throats, eyes, ears, and the nose if the chemical is inhaled. Concentrated levels indoors can also be flammable. If possible, avoid products containing formaldehyde.

- Ammonia: Exposure to this chemical can cause lung damage, blindness, and death. Just ingesting a tiny amount will cause the throat to feel like it's burning. The chemical itself is natural, the body uses ammonia to make proteins and other molecules, but in large artificial doses, it poses a real threat.

- Sodium Lauryl Sulphate: Exposure to this chemical can cause immunological problems including cancer while a small exposure may cause skin, hair, and eye damage.

Healthier Alternatives for Home and Environment

It is possible to clean your home in a green way while still making it safe to live in; even for children. To do this, you need to consider what natural products there are out there which can also be used for cleaning items. For example, surfaces, floors, teddy bears, windows, and so on. Here are a few suggestions:

White Vinegar: When mixed with warm water, and if you prefer a few drops of tea tree oil, this white vinegar homemade mixture can be used to clean almost any surface.

Lemons: Just like with vinegar, lemon juice mixed with warm water will clean surfaces and objects. The leftover halves of a lemon are also good for getting rid of bad smells.

In addition to this, you can use baking powder mixed with water and white vinegar to clean many things such as pots, pans, kettles and messy ovens. With a little research, you can find many good, edible products which are good for the environment.

Environmental Impact of Concrete

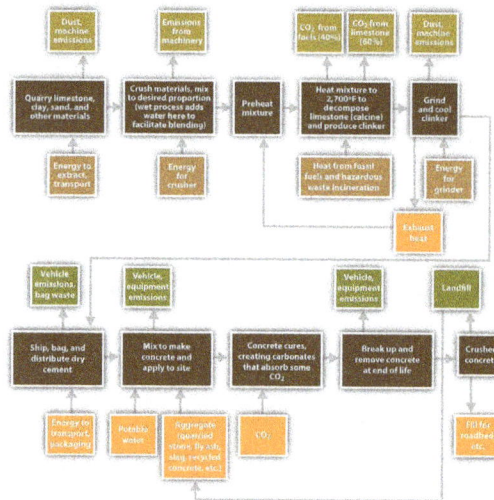

Cement and Global Warming

- Making cement results in high levels of CO_2 output.

- Cement production is the third ranking producer of anthropogenic (man-made) CO_2 in the world after transport and energy generation.

- 4 - 5% of the worldwide total of CO_2 emissions is caused by cement production.

- CO_2 is produced at two points during cement production:

  o the first is as a byproduct of burning of fossil fuels, primarily coal, to generate the heat necessary to drive the cement-making process.

  o the second from the thermal decomposition of calcium carbonate in the process of producing cement clinker.

- $CaCO_3$ (limestone) + heat -> CaO (lime) + CO_2.

- Production of one tonne of cement results in 780 kg of CO_2.

- Of the total CO_2 output, 30% derives from the use of energy and 70% results from decarbonation.

- Important to realise is that although 5% of the worldwide generation of CO_2 is due to cement production, that level of output also reflects the unique and universal importance of concrete throughout the construction industry.

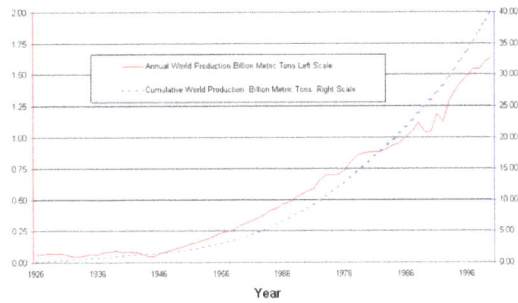

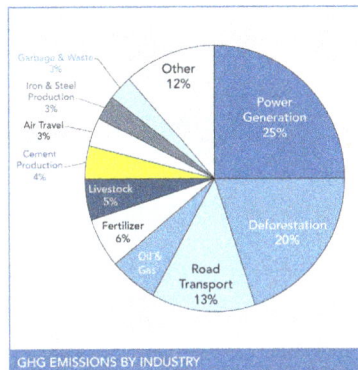

GHG EMISSIONS BY INDUSTRY

Concrete Raw Materials Extraction and Processing

- Landscape degradation
- Dust and noise
- Visual impact on some areas of outstanding natural beauty
- Proximity to population centers
- Loss of agricultural land
- Use of potable water to wash aggregates, dust suppression and in the manufacturing process

Transport

- Energy consumption
- Vehicle pollution
- Noise

Methods to Reduce Environmental Impact of Concrete

Following are the various methods which can considered to reduce the environmental impact of concrete:

- Cement conservation
- Aggregate conservation
- Water conservation
- Concrete durability

Cement Conservation

The conservation of cement is the first and most important step in decreasing both energy utilization and greenhouse gas emission. Resource productivity consideration stipulates the deduction of the utilization of Portland cement while meeting the future demands for more concrete. This needs to be the first agenda of successful concrete industry.

The use of Portland cement containing pozzolanic materials for example ground granulated blast furnace slag, fly ash, and silica fume is increasing considerably. However, these by product admixture which are used as cement replacement material are employed in minor or low value applications for instance landfills and road sub-bases.

The use of cement in structural applications can be decreased by using by product cementitious or pozzolanic material as a cement replacement. This leads to decrease the demand for cement production.

It is reported that, replacing cement with slag or fly ash by 50% will provide better durable product compare with that of Portland cement with zero replacement and consequently natural resource application is decreased.

Furthermore, the setting and hardening of concrete containing large percent of cementitious material is low but this can be tackled to a certain extent by using superplasticizer.

It is possible that slow pace construction process be approved in the future when resource maximization is become the most important industry purpose instead of labor productivity.

Aggregate Conservation

It is claimed that in North America, Japan, and Europe, around two thirds of construction and demolition waste are composed of old broken concrete and masonry. If these waste materials are reused as a coarse aggregate, material productivity will improve greatly.

Moreover, dredge sand and mining waste which are present in number of countries around the world, can be processed and applied as fine aggregate. Even though the processing of these waste material require a budget but it can be considerably economical particularly in those countries where the cost of damping waste material is substantial and land is rare.

So significant is the recycling and reusing waste material that solve natural resource depletion problem in many regions and avoid high cost of transporting virgin aggregate over long distances.

It is claimed by Lauritzen that, 1 billion tons of concrete and masonry rubble are produced per year and small amount of concrete and masonry waste are reused again.

Expensive waste disposal and environmental considerations have motivated the majority of European country to set short term goals to recycle between 50 to 90% of demolition and construction waste.

Lastly, recycled aggregate in general and specifically masonry aggregate possess large porosity compare with natural aggregate. So, for the same workability, water demand to produce fresh concrete is higher compare the case of using natural aggregate and the mechanical properties of hardened concrete are influenced detrimentally. To tackle this problem, combination of natural and recycle aggregate may be used or fly ash and water reducing admixture can be employed in concrete.

Water Conservation

It is reported by Hawken et al that, the availability of fresh and clean water decreases continuously and only 3% of all water on earth are fresh water which most of it either located deep beneath earth surface of trapped fast melting glaciers.

As a result of increasing industrial, agricultural, and urban demand for water, water table is lowering in addition to the increase of water contamination.

It is recommended that, the only practical and reasonable solution to this problem is the utilization of available resources more efficiently.

Concrete producers consume water in large scale and these producers and other fresh large consumers should be forced to use water efficiently. It is estimated that $100L/m3$ is used to clean ready mix trucks and large amount of water is employed for mixing.

It is believed that, 1 trillion L of water is used for mixing annually and this huge quantity can be decreased to half by the increase of mineral admixture and superplasticizer application and better grading of aggregate.

Moreover, with the approval of test results, the use of brackish water and industrial recycled water must be enforced instead of clean water, and this must be entirely imperative in the case of washing equipments.

Furthermore, it is reported that considerable amount of water saved when retarder used for fresh returned concrete.

Finally, during concrete curing, the application of textile, which has exterior impermeable membrane and water absorbent fabric at interior face, cut water utilization.

Concrete Durability

In addition to those measures discussed in the above sections, enhancing concrete durability offer long term solution and exceptional breakthrough for improving productivity of concrete industry and hence decrease environmental impact on concrete production, for instance if concrete structure is constructed for a service life of 500 year rather than 50 year, the resource productivity of concrete industry will increased by factor of 10.

The durability of modern structures is questionable because deterioration begins after around 20 years whereas there are buildings and seawalls constructed from unreinforced roman concrete which maintain their good condition after nearly 2000 years. This might mainly because of considerably crack prone Portland cement concrete which consequently became permeable during its service life.

Moreover, steel reinforcement in permeable concrete corrodes and leads to progressive damage of the structure.

In modern times, construction practice is controlled by culture of accelerating construction speed in which large amount of high early strength Portland cement is employed. Consequently, weak crack resistance concrete structure is constructed because of large drying shrinkage and thermal contraction and small creep relaxation.

Furthermore, roman concrete made with mixture of volcanic ash and hydrated lime, and created homogenous hydrated product that set and hardened in a slow pace however better than hydrated Portland cement product thermodynamically. Added to that, less amount of water used in the roman concrete and were not subjected to cracks at the same degree as Portland cement concrete.

Therefore, if concrete durability is major concern or purpose, producing less crack-prone concrete rather than high speed construction should be focused on and hence construction practice need to undergo paradigm change toward that direction.

It is demonstrated that, entire of most of cracking and shrinkage in concrete can be prevented and consequently high durable concrete can be produced provided that water to cementitious material in concrete is deduced through superplasticizer application.

A large free crack monolithic concrete foundation of a temple that is in Kauai which is an island in Pacific Ocean is described by Mehta and Langley. The foundation made of two parallel of unreinforced concrete slab.

To produce concrete with considerably small shrinkage stresses, it was necessary to decrease to limit both thermal and drying shrinkage through substantial reduction of Portland cement and water in concrete.

Slump of concrete used for the foundation was 125±25 mm and its compressive strength after 90 day was 20 MPa with 13 C rise of temperature. The exposed surface of the foundation was carefully and properly examine after almost two year and there was no any crack evidence.

Investigation on the core samples taken from the slab demonstrated that, not only did the homogeneity of the hydration product of high fly ash system were better compare with conventional concrete but also the bond between aggregate and hydration product was very well. This is a precondition for achieving crack resistant and high durable concrete.

Figure shows thin section of core sample taken from the slab and neither interfacial zone of micro-cracking are shown between coarse aggregate and adjacent cement mortar.

In contrary, Figure illustrates the how interfacial aggregate-paste micro-cracks connect and allow the ingression of fluid from outside

Finally, the mixture proportion employed to produce concrete in the construction of foundation of the temple in Kauai are shown in table below.

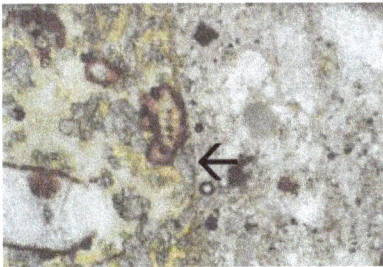

| Photomicrograph of Thin Section from Concrete Core obtained from High Volume Fly Ash System | Photomicrograph of a Thin Section taken from Conventional Portland Cement |

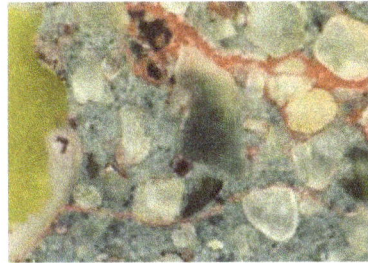

Mixture Proportions of Crack Resistant High Volume Fly Ash Concrete

Constituents of the mixture	Proportions by weight (Kg/m³)
Type I Portland cement	106
Class F fly ash	142
Water	100
Crushed calcareous sand	944
Crushed basalt rock 25mm maximum size	1120
Superplasticizer	3.5 L/m3

Environmental Management

Environmental management consists of organizing different environmental initiatives to address various ecological issues that are affecting the globe. Environmental management deals with trying to prevent ecological disaster as well as aiding in environmental crises and trying to find appropriate solutions. Environmental management looks at land, marine and atmospheric conditions, such as global warming, marine-life preservation and deforestation.

The main aim of environmental management is to ensure that we leave the planet in a healthy state for future generations and to help preserve all forms of life, including marine life and vegetation. To do so, environmental managers need to consider the carbon footprint of the current human generation and look at ways to minimize any irreversible damage we are leaving behind. One of the main studies of environmental managers is investigating possible renewable sources of energy, to ensure there is enough fossil fuel, which takes millions of years to regenerate, left once our generation is gone.

The process, environmental management is related to the rational adjustment of man with nature involving judicious exploitation and utilization of natural resources without disturbing the ecosystem balance and ecosystem equilibrium.

If the natural resources are overexploited, it will affect socio-economic development of a nation. Thus, environmental management must take into consideration the ecological principles and socioeconomic needs of the society i.e., it involves socio economic developments on one hand and maintenance of environmental quality on other hand.

It is clear that environmental managements have two major aspects:

  i)   Socio-economic development. and

  ii) Stability of biosphere in general and stability of individual ecosystems in particular.

Scope and Aspects of Environmental Management

Environmental management is very wide in scope and includes all the technical, economical and other aspects of environment.

The broader objectives of environmental management includes:

  i)   To identify the environmental problem and to find its solution.

  ii) To restrict and regulate the exploitation and utilization of natural resources.

  iii) To regenerate degraded environment and to renew natural resources (renewable).

  iv) To control environmental pollution and gradation.

  v)   To reduce the impacts of extreme events and natural disaster.

  vi) To make optimum utilization of natural resources.

  vii) To assess the impacts of proposed projects and activities on environment.

  viii) To review and revise the existing technologies and make them ecofriendly.

  ix) To formulate laws for the implementation of environmental protection and conservation programmes.

The components of environmental management are based on five fundamental aspects.

1. Environmental perception and public awareness

The environmental perception and public awareness considers the following points:

   a) Sources of environmental perception and public awareness.

   b) Level of environmental perception.

   c) Role of environmental perception in environmental planning and management.

2. Environmental education and training

Environmental education and training should be given at school, college and University levels by professionals.

3. Resource management

The resource management considers the following points:

   i) Classification of natural resources.

   ii) Survey and evaluation of ecological resources.

   iii) Preservation of resources.

   iv) Conservation of resources.

4. Control of Environmental degradation and pollution

The environmental degradation and pollution can be checked by considering the following points:

   i) Control of environmental degradation and pollution.

   ii) Adopting suitable preventive mechanisms to reduce natural hazards and disaster.

   iii) Regeneration of degraded environment.

5. Environmental impact assessment

The environmental impact assessment involves:

   i) Appraisal of existing environmental conditions.

   ii) Appraisal of existing and proposed production methods.

   iii) Methologies and procedures.

   iv) Probable impacts of existing and proposed project.

   v) Review of technology and required improvement.

Water Resource Management

Water resource management encompasses a wide range of disciplines and expertise, including integrated water resources management, assessment of the demand for water for drinking and irrigation purposes, groundwater mapping and surface water monitoring, simulation and modelling of the hydrologic cycle, legislation, river basin management, institutional and capacity building and water resources protection.

An Integrated Approach to Water Management

The key challenges of contemporary water management can only be understood within the very broad context of the world's socio economic systems. It is widely accepted that sustainable and equitable water management only can be achieved using an integrated approach. Assessment of the resource is the basis for rational decision-making, and authorities that use such assessments must be further supported and expanded from local to international levels.

Ensuring Safe Drinking Water

Access to safe drinking water is a basic human right and a key component of an efficient policy for health protection. Managing water resources to ensure satisfying drinking water quality is a global concern and a priority for sustainable development - the supply of drinking water of good quality is the basis for the proper functioning society.

Surface water is used as the primary drinking water resource in most countries.

Uneven distribution of surface water and the deterioration in quality is likely to result in an increasing reliance on groundwater resources during the coming decades. Increasing industrialization and thereby increasing risk of pollution is added to the risks of over-abstraction of groundwater, potential given subsidence or saline intrusion. When combining surface and groundwater, an integrated approach to water resources management will become ever more important.

Reuse of water has within the last decade been introduced many places as a new or additional source of water.

Solid Waste Management

Garbage arising from human or animal activities, that is abandoned as unwanted and useless is referred as solid waste. Generally, it is generated from industrial, residential and commercial activities in a given area, and may be handled in a variety of ways. However, waste can be categorized based on materials such as paper, plastic, glass, metal and organic waste. Solid waste disposal must be managed systematically to ensure environmental best practices. Solid waste disposal and management is a critical aspect of environmental hygiene and it needs to be incorporated into environmental planning.

Solid waste disposal and management includes planning, administrative, financial, engineering and legal functions. It is typically the job of the generator, subject to local, national and even international authorities.

Solid waste disposal management is usually referred to the process of collecting and treating solid wastes. It provides solutions for recycling items that do not belong to garbage or trash. Solid waste management can be described as how solid waste can be changed and used as a valuable resource.

Improper disposal of municipal solid waste can create unsanitary conditions, and these conditions in turn lead to pollution of the environment. Diseases can be spread by rodents and insects. The tasks of solid waste disposal management are complex technical challenges. They can also pose a wide variety of economic, administrative and social problems that must be changed and solved.

Many broad categories of garbage are:

i.   Organic waste: kitchen waste, vegetables, flowers, leaves, fruits.

ii. Toxic waste: old medicines, paints, chemicals, bulbs, spray cans, fertilizer and pesticide containers, batteries, shoe polish.

iii. Recyclable: paper, glass, metals, plastics.

iv. Hospital waste such as cloth with blood.

Effects of Solid Waste Pollution

Municipal solid wastes heap up on the roads due to improper disposal system. People clean their own houses and litter their immediate surroundings which affects the community including themselves.

This type of dumping allows biodegradable materials to decompose under uncontrolled and unhygienic conditions. This produces foul smell and breeds various types of insects and infectious organisms besides spoiling the aesthetics of the site. Industrial solid wastes are sources of toxic metals and hazardous wastes, which may spread on land and can cause changes in physicochemical and biological characteristics thereby affecting productivity of soils.

Toxic substances may leach or percolate to contaminate the ground water. In refuse mixing, the hazardous wastes are mixed with garbage and other combustible wastes. This makes segregation and disposal all the more difficult and risky.

Various types of wastes like cans, pesticides, cleaning solvents, batteries (zinc, lead or mercury), radioactive materials, plastics and e-waste are mixed up with paper, scraps and other non-toxic materials which could be recycled. Burning of some of these materials produces dioxins, furans and polychlorinated biphenyls, which have the potential to cause various types of ailments including cancer.

Methods of Solid Wastes Disposal

i.   Sanitary Landfill

ii. Incineration

iii. Composting

iv. Pyrolysis

Sanitary Land Filling

In a sanitary landfill, garbage is spread out in thin layers, compacted and covered with clay or plastic foam. In the modern landfills the bottom is covered with an impermeable liner, usually several layers of clay, thick plastic and sand. The liner protects the ground water from being contaminated due to percolation of leachate.

Leachate from bottom is pumped and sent for treatment. When landfill is full it is covered with clay, sand, gravel and top soil to prevent seepage of water. Several wells are drilled near the landfill site to monitor if any leakage is contaminating ground water. Methane produced by anaerobic decomposition is collected and burnt to produce electricity or heat. Sanitary Landfills Site Selection:

i.   Should be above the water table, to minimize interaction with groundwater.

ii. Preferably located in clay or silt.

iii. Do not want to place in a rock quarry, as water can leech through the cracks inherent in rocks into a water fracture system.

iv. Do not want to locate in sand or gravel pits, as these have high leeching. Unfortunately, most of Long Island is sand or gravel, and many landfills are located in gravel pits, after they were no longer being used.

v.   Do not want to locate in a flood plain. Most garbage tends to be less dense than water, so if the area of the landfill floods, the garbage will float to the top and wash away downstream.

A large number of adverse impacts may occur from landfill operations. These impacts can vary:

i.   Fatal accidents (e.g., scavengers buried under waste piles).

ii. Infrastructure damage (e.g., damage to access roads by heavy vehicles).

iii. Pollution of the local environment (such as contamination of groundwater and/or aquifers by leakage and residual soil contamination during landfill usage, as well as after landfill closure).

iv. Off gassing of methane generated by decaying organic wastes (methane is a greenhouse gas many times more potent than carbon dioxide, and can itself be a danger to inhabitants of an area).

v.   Harbouring of disease vectors such as rats and flies, particularly from improperly operated landfills.

Incineration

The term incinerates means to burn something until nothing is left but ashes. An incinerator is a unit or facility used to burn trash and other types of waste until it is reduced to ash. An incinerator is constructed of heavy, well-insulated materials, so that it does not give off extreme amounts of external heat.

The high levels of heat are kept inside the furnace or unit so that the waste is burned quickly and efficiently. If the heat were allowed to escape, the waste would not burn as completely or as rapidly. Incineration is a disposal method in which solid organic wastes are subjected to combustion so as to convert them into residue and gaseous products. This method is useful for disposal of residue of both solid waste management and solid residue from waste water management. This process reduces the volumes of solid waste to 20 to 30 per cent of the original volume.

Incineration and other high temperature waste treatment systems are sometimes described as "thermal treatment". Incinerators convert waste materials into heat, gas, steam and ash. Incineration is carried out both on a small scale by individuals and on a large scale by industry. It is used to dispose of solid, liquid and gaseous waste. It is recognized as a practical method of disposing of certain hazardous waste materials. Incineration is a controversial method of waste disposal, due to issues such as emission of gaseous pollutants.

Composting

Due to shortage of space for landfill in bigger cities, the biodegradable yard waste (kept separate from the municipal waste) is allowed to degrade or decompose in a medium. A good quality nutrient rich and environmental friendly manure is formed which improves the soil conditions and fertility.

Organic matter constitutes 35%-40% of the municipal solid waste generated in India. This waste can be recycled by the method of composting, one of the oldest forms of disposal. It is the natural process of decomposition of organic waste that yields manure or compost, which is very rich in nutrients.

Composting is a biological process in which micro-organisms, mainly fungi and bacteria, convert degradable organic waste into humus like substance. This finished product, which looks like soil, is high in carbon and nitrogen and is an excellent medium for growing plants.

The process of composting ensures the waste that is produced in the kitchens is not carelessly thrown and left to rot. It recycles the nutrients and returns them to the soil as nutrients. Apart from being clean, cheap, and safe, composting can significantly reduce the amount of disposable garbage.

The organic fertilizer can be used instead of chemical fertilizers and is better specially when used for vegetables. It increases the soil's ability to hold water and makes the soil easier to cultivate. It helped the soil retain more of the plant nutrients.

Vermi-composting has become very popular in the last few years. In this method, worms are added to the compost. These help to break the waste and the added excreta of the worms makes the compost very rich in nutrients. In the you can learn how to make a compost pit or a vermi-compost pit in your school or in the garden at home.

To make a compost pit, you have to select a cool, shaded corner of the garden or the school compound and dig a pit, which ideally should be 3 feet deep. This depth is convenient for aerobic composting as the compost has to be turned at regular intervals in this process.

Preferably the pit should be lined with granite or brick to prevent nitrite pollution of the subsoil water, which is known to be highly toxic. Each time organic matter is added to the pit it should be covered with a layer of dried leaves or a thin layer of soil which allows air to enter the pit thereby preventing bad odor. At the end of 45 days, the rich pure organic matter is ready to be used. Composting: some benefits:

   i.   Compost allows the soil to retain more plant nutrients over a longer period.

   ii. It supplies part of the 16 essential elements needed by the plants.

iii. It helps reduce the adverse effects of excessive alkalinity, acidity, or the excessive use of chemical fertilizer.

iv. It makes soil easier to cultivate.

v. It helps keep the soil cool in summer and warm in winter.

vi. It aids in preventing soil erosion by keeping the soil covered.

vii. It helps in controlling the growth of weeds in the garden.

Pyrolysis

Pyrolysis is a form of incineration that chemically decomposes organic materials by heat in the absence of oxygen. Pyrolysis typically occurs under pressure and at operating temperatures above 430 °C (800 °F).

In practice, it is not possible to achieve a completely oxygen-free atmosphere. Because some oxygen is present in any pyrolysis system, a small amount of oxidation occurs. If volatile or semi-volatile materials are present in the waste, thermal desorption will also occur.

Organic materials are transformed into gases, small quantities of liquid, and a solid residue containing carbon and ash. The off-gases may also be treated in a secondary thermal oxidation unit. Particulate removal equipment is also required. Several types of pyrolysis units are available, including the rotary kiln, rotary hearth furnace, and fluidized bed furnace. These units are similar to incinerators except that they operate at lower temperatures and with less air supply.

Limitations and Concerns

i.   The technology requires drying of soil prior to treatment.

ii. Limited performance data are available for systems treating hazardous wastes containing polychlorinated biphenyls (PCBs), dioxins, and other organics. There is concern that systems that destroy chlorinated organic molecules by heat have the potential to create products of incomplete combustion, including dioxins and furans. These compounds are extremely toxic in the parts per trillion ranges. The MSO process reportedly does not produce dioxins and furans.

iii. The molten salt is usually recycled in the reactor chamber. However, depending on the waste treated (especially inorganics) and the amount of ash, spent molten salt may be hazardous and require special care in disposal.

iv. Pyrolysis is not effective in either destroying or physically separating in organics from the contaminated medium. Volatile metals may be removed as a result of the higher temperatures associated with the process, but they are not destroyed. By-products containing heavy metals may require stabilization before final disposal.

v.   When the off-gases are cooled, liquids condense, producing an oil/tar residue and contaminated water. These oils and tars may be hazardous wastes, requiring proper treatment, storage, and disposal.

Air Resource Management

The word increase in the use of the air resource management concept is gratifying to observe. Much is being said today about the need to manage our air resource, and it is timely to define these terms explicitly.

In the scheme of things, the basic use of our air resource is to sustain life. All other uses must yield to the maintenance of air quality that will not degrade, either acutely or chronically, the health or well-being of man.

Two major areas of possible adjustment exist: the aesthetics and the economic impact of air pollution and its control. The cost borne by society to achieve a desired quality of air should, therefore, be in balance with the benefits to be attained.

The proper management of our air resources must, for the present and future, consider these major tenets.

Limitations of the Air Resource

Our atmosphere extends far above the earth's surface, the upper limit being impossible to define exactly. Some 95% of the total air mass, however is concentrated in a comparatively thin layer that extends some 12 miles above the crust of the earth. Totally, we are richly endowed with an atmospheric resource of great magnitude, and most of it is located proximate to the earth's surface, where it is of most utility to man and his activities.

What is important to us, however, is not the total volume of our atmosphere, but the rate at which air is available for use at specific locations. Our use of the limited available supply should not alter its physical or chemical composition to interfere with its subsequent physiological use agriculturally, domestically, or industrially, or otherwise produce adverse or undesirable effects in the environment. Most of this discussion will deal with the "judicious use of means" to mitigate the adverse effects from man's pollution of the atmosphere. This challenge will be examined from a local or regional view with some mention of the role of the federal government in developing the air management concept.

Elements of Air Resource Management Programs

Some of the elements of an air re source management program are:

1. Development of a public policy on air conservation.

2. An organizational framework and 8taff capable of operating along functional lines (e.g., engineering, technical services, field services), and the funding support.

3. Delineation of realistic short range goals that can be effectively met within a reasonable time, i.e., 5 years.

4. Assessment, continuously, of existing air quality and preparation of esti mates of the future situation.

5. Assessment in depth on a continuing basis, of the emissions from all existing and future pollution sources.

6. Development of the necessary Information about factors that influence the transport of air pollutants.

7. Assessment of ambient air quality of a community or region on man and his environment.

8. Establishment of ambient air quality goals (referred to by some as objectives or standards).

9. Design of remedial measures and programs calculated to bring about the air quality desired.

10. Development of long-range air use plans that are fully integrated with other community plans for land use, transportation, recreation, refuse disposal, etc., to cope effectively with projected changes for the community.

11. Development of an understanding of the broad implications of science and technology on air resources and the potential effect on the social and economic character of the modern society.

12. Development of an effective in formation and educational program to inform the community of present and future problems, and of the need to solve them promptly and effectively.

Urban Air Resource Management Programs

The mechanisms and organization for reaching decisions in a community (or state) are of key importance in the success of an air resource management program (or any other community effort). These mechanisms must pro vide for the assimilation of information into goals and policies and the sub sequent enunciation of a public policy on the management of a community's air resource. Appropriate community involvement in the decision making process must be encouraged to ensure a base of public support for action.

The degree of success achieved in making decisions and developing policy will depend upon the clear delegation of responsibility to a key individual. It follows that full support by top management is needed to develop and implement a set of program goals. There is no doubt that the situation may be complex, but is certainly amenable to solution.

Long-range planning must be an integral part of an air resource management program. The way is smoothed for anticipated social and technological changes in the urban environment and it is possible to consider a much broader scope of alternative programs to achieve or maintain high-quality air. For example, as one means of reducing pollutant emissions, in a 25-year plan, consideration could be given to eliminating certain existing combustion operations and substituting different sources of energy as a means for heating. Such a program would be too complicated to plan for on a short-range basis. Planning also forces program directors to think ahead about technological changes and predisposes them toward changes that they can agree to and support financially. Without planning, the director will find himself reacting to changing events rather than influencing them as he should.

References

- Effects-environmental-pollution: abcofagri.com, Retrieved 11 June 2018

- Nuclear-energy-affect-environment-4566966: sciencing.com, Retrieved 15 March 2018

- Environmental-impact-cleaning-product: greenblizzard.com, Retrieved 19 July 2018

- Environmental-impacts-of-concrete, building-design: greenspec.co.uk, Retrieved 10 April 2018

- Types-of-solid-waste-disposal-and-management: pulpandpaper-technology.com, Retrieved 13 June 2018

- Solid-waste-management-types-sources-effects-and-methods-of-solid-waste-management-9949: yourarticlelibrary.com, Retrieved 16 April 2018

Permissions

Index